University of Hertfordshire

Learning and Information Services

Hatfield Campus Learning Resources Centre
College Lane Hatfield Herts AL10 9AB
Renewals: Tel 01707 284673 Mon-Fri 12 noon-8pm only

This book is in heavy demand and is due back strictly by the last date stamped below. A fine will be charged for the late return of items.

ONE WEEK LOAN

PRACTICAL EXTRUSION BLOW MOLDING

PLASTICS ENGINEERING

Founding Editor

Donald E. Hudgin

Professor
Clemson University
Clemson, South Carolina

Additional Volumes in Preparation

PRACTICAL EXTRUSION BLOW MOLDING

edited by

SAMUEL L. BELCHER

Sabel Plastechs, Inc.
Moscow, Ohio

MARCEL DEKKER, INC.   NEW YORK · BASEL

Cover photo, twin-head parison extrusion for "F" style container, courtesy of Johnson Controls, Inc., Plastic Machinery Division, Manchester, MI.

ISBN: 0-8247-1997-2

This book is printed on acid-free paper.

Headquarters
Marcel Dekker, Inc.
270 Madison Avenue, New York, NY 10016
tel: 212-696-9000; fax: 212-685-4540

Eastern Hemisphere Distribution
Marcel Dekker AG
Hutgasse 4, Postfach 812, CH-4001 Basel, Switzerland
tel: 41-61-261-8482; fax: 41-61-261-8896

World Wide Web
http://www.dekker.com

The publisher offers discounts on this book when ordered in bulk quantities. For more information, write to Special Sales/Professional Marketing at the headquarters address above.

Current printing (last digit):
10 9 8 7 6 5 4 3 2 1

PRINTED IN THE UNITED STATES OF AMERICA

PREFACE

Today, when walking through your supermarket, it is increasingly difficult to find items packaged in glass bottles and jars. Packaging for soft drinks, healthcare and beauty products, household chemicals and over-the-counter medicines, among other products, have switched from glass or metal to plastic. The soft drinks market went from zero plastic containers in 1977 to more than 10 billion in 1999. In the United States in 1997, the blow molding industry produced over 20 billion plastic containers and consumed over 3.5 billion pounds of resin.

The blow molding industry has expanded over the past 40 years from simple plastic containers to plastic drums, gas tanks, storage tanks, automobile bumpers, automobile spoilers, furniture, and, naturally, toys of all shapes and sizes.

Training in the extrusion blow molding industry is one of today's greatest challenges. Changes in product design through the use of computers and rapid prototyping have meant that consumer items are produced via extrusion blow molding that just three years ago would have been considered impractical. New resins and faster and larger machines with smart controls place an even greater burden than ever on the production floor personnel, and this trend will continue.

This book focuses on extrusion blow molding to provide a single source for guidance and training in everyday operations for personnel at all levels of production.

The material for this book was contributed by people in industry who have given their time to share their knowledge—in layman's terms—so others may learn from their experiences. My dad always said, "If you want something done, give it to a busy person." I never understood this statement until I too was in industry. These busy people made this book possible. To all the contributors, their companies that have supported them, and their families, and to companies who have granted permission to use their data and photographs, I thank you. A special thanks to Linda Schonberg, Production Editor, and to Marcel Dekker, Inc., for assisting in bringing this book to the industry.

Samuel L. Belcher

CONTENTS

CONTRIBUTORS

Samuel L. Belcher President, Sabel Plastechs, Inc., Moscow, Ohio

Charles Carey Vice President of Sales/Marketing, Engineering, and Manufacturing, Wentworth Mould and Die Company Ltd., Hamilton, Ontario, Canada

Cheryl L. Hayek TopWave International, Inc., Marietta, Georgia

Robert R. Jackson Jackson Machinery, Inc., Port Washington, Wisconsin

Reinhold Nitsche Engineering Manager, Wentworth Mould and Die Company Ltd., Hamilton, Ontario, Canada

Alex Orlowsky   Seajay Manufacturing Corporation, Neptune, New Jersey

Anatoly Orlowsky   Seajay Manufacturing Corporation, Neptune, New Jersey

Don L. Peters   Principal Engineer, Blow Molding, Plastics Technical Center, Phillips Chemical Company, Bartlesville, Oklahoma

Robert J. Pierce   Assistant Professor, College of Technology, Ferris State University, Big Rapids, Michigan

J. L. Rathman   Blow Molding Specialist, Plastics Technical Center, Phillips Chemical Company, Bartlesville, Oklahoma

Karl Schwarze   Senior Mould Designer, Wentworth Mould and Die Company Ltd., Hamilton, Ontario, Canada

Robert A. Slawska   Director, Proven Technology, Inc., Belle Mead, New Jersey

W. Bruce Thompson   Molding Manager, Fluoroware, Inc., Chaska, Minnesota

Tar Tsau   Engineering Manager, Engineering Department, Wentworth Mould and Die Company Ltd., Hamilton, Ontario, Canada

Dan Weissmann*   Director, Development and Application Engineering, Schmalbach-Lubeca Plastic Containers USA, Inc., Manchester, Michigan

---

**Current affiliation*: Consultant, Simsbury, Connecticut.

PRACTICAL
EXTRUSION
BLOW
MOLDING

1

HISTORY OF BLOW MOLDING

Samuel L. Belcher
Sabel Plastechs, Inc.
Moscow, Ohio

1.1 INTRODUCTION

Extrusion blow molding is a technique for producing a hollow product. This is accomplished by forming a heated tube of a thermoplastic material called either a parison or preform and placing it in a blow mold containing a female cavity. Once the parison is in the blow mold and it is closed on the heated thermoplastic parison, the parison is inflated with a gas, usually air, that blows the heated parison to take the form of the mold cavity. The mold allows the heated thermoplastic material to cool, exhausts the gas, and then the mold is opened to extract the formed plastic product.

The main advantage of blow molding is its ability to produce varied designs of hollow thermoplastic articles. Extru-

Courtesy of Johnson Controls, Inc., Plastics Machinery Division, Manchester, MI.

sion blow molding offers flexibility of article design, handle ware, and intricate shapes with the many different thermoplastics that are in the marketplace today to satisfy customer requirements.

Extrusion blow molded thermoplastic containers are today the choice of the consumer. Extrusion blow molded articles are lightweight, safe, recyclable, low cost, easy to handling, water clear or color enhanced, and easy to use (Fig. 1.1).

1.2 HISTORY

Credit for originating the blow molding industry in the 1930s is given to the Hartford Empire Company. However there is evidence that the Egyptians and Babylonians blew plastic materials into various utensils. Synthetic plastics were blown commercially in the 1880s when celluloid was blown into items, such as baby rattles, hair brush backs, ping-pong balls, doll parts, cutlery handles, and jewelry. Methods involved

Figure 1.1 Courtesy of Johnson Controls, Inc., Plastics Machinery Division, Manchester, MI.

softening sheets with heat or immersion in hot liquids followed by placement in crude molds and blowing with air. Hollow articles made from celluloid and cellulose acetate continued to be produced by these methods well into the 1930s.

Enoch T. Ferngren is credited with being the first to make blown plastic items in the "modern era" of blow molding. Mr. Ferngren along with William Kopitke, designed and produced a mechanical device to produce blown plastic articles and sold it to the Hartford Empire Company in 1937. In 1940 the Plax Corporation was formed for blow molding and in 1940 produced hollow Christmas tree balls from cellulose acetate. Owens Illinois Glass Company was experimenting with all available plastics on a modified ram injection machine and produced 20 cc bottles via the injection blow molding process. Owens Illinois Glass company supplied canteens and 100 cc polystyrene bottles to the United States Army Medical Corps during World War II. Only a few products could be packaged in plastics in the 1940s due to the poor barrier properties of the plastics plus the flavor and taste transmission.

From the 1930s and well into the 1940s, many innovations were tried to allow blow molding the plastics available then. Hydraulics were being used for simple motions, and the more complicated motions were done by mechanical means. Steam and crude electrical elements were tried as heating media. The plastic materials used were cellulose acetate, polystyrene and some polyvinyl chloride. The industry finally blossomed into reality when two things took place in 1945. Low density polyethylene (LDPE) was patented in 1937, and the first commercial plant went on stream in 1939. The first available LDPE was used for the war effort. However, in 1945, a small quantity became available for commercial use (Fig. 1.2). Plax introduced the "Stopette" deodorant squeeze bottle in 1945, and over five million units were sold. The industry had its first real commercial success.

Until this time all the machinery for blow molding was developed by the blow molders. The success of the "Stopette" squeeze bottle opened the door, and companies that built rubber machines and injection machines began in earnest to design and build blow molding machines. However, only small items were considered possible to blow mold until high density polyethylene (HDPE) became commercially available to the industry and particularly for blow molding in 1956.

In 1957 there were only three blow molders listed in the SPI Directory and in 1962 Directory this had grown only to thirteen which could produce bottles.

In 1945 Mr. James Bailey authored an article describing indirect, direct, and diaphragm methods that had been investigated and stated that the future of blow molding would be in extrusion and injection blow molding.

There were many inventions and improvements in the entire plastics industry from 1945 to 1950. However, the major advances were made from 1950 into the early 60s. Com-

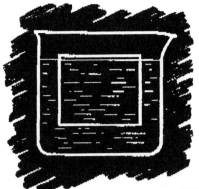

Figure 1.2 1945 advertisement for a polyethylene made by DuPont.

mercial blow molding machines were being offered by Bekum, Kautex, and Fischer, and Continental Can had developed the Mills wheel (see Tables 1.1 and 1.2).

Once HDPE became commercially available in 1956, end users felt they had a rigid plastic that could replace other ma-

Table 1.1

Blow Molding: 1930–1994

Equipment Development

1930	Fernplas Corporation blows cellulose acetate with a machine
1935	Ferngren patent for extruding a molten tube into a closed mold and inject air
1935	First modern injection machine imported into USA
1937	Ferngren and Kopitke make and demonstrate first mechanical device to blow mold. Ram injection used.
1938	BASF extrusion blow molds in Europe
1939	Plax operates blow molding machines to make 25,000 blow molded items per day
1939	Electrically heated extruder developed
1942	Owens-Illinois patents an injection blow process with core pins

Other Developments Important to Blow Molding

Existing plastic processing machine modified for blow molding. Specific designs for blow molding

Only ram injection machines available. Extrusion well established.

Injection parison and shift to blow mold developed. Tube cutting, needle blow, hollow tube extrusion, tube transfer to blow mold, pinch tube to make parison, and other basic operations resolved.

Available Plastic Materials

Cellulose acetate (1927), PVC (1927), acrylics (1936), cellulose acetate butyrate (1938), nylon (1935), polyvinylidene chloride (1939), polystyrene (1938), cellulose propionate (1945), ethyl cellulose (1936), experimental LDPE (1942), polyesters (1942)

Blow Molded Products

Christmas tree balls (1939), hollow ware, toys, brush backs, baby rattles, ping-pong balls, small bottles, canteens, toilet floats, novelties

terials, such as corrugated cardboard, glass, paper, molded pulp, and the original plastics, such as PVC, polystyrene, LDPE, and cellulose acetate.

In the early 1960s other plastics were being developed, such as polycarbonate, acrylics, polyacrylic nitrile, phenoxy, polypropylene, nylon, and polyacetals. These new materials offered clarity, rigidity, grease resistance, less permeation by certain gases, chemical resistance, and enabled plastic bottle producers to become competitive with glass, metal, and composites.

In 1964 a comprehensive summary was published of the six methods that are basic to much that has since been developed: injection-transfer, hand transfer, fixed mold with interrupted extrusion, alternate flow using manifold extrusion, extrusion, and using molds with continuous extrusion plus the horizontal or vertical multistation rotary.

In 1961 an article was published on how to design heads, dies, and molds plus information on flow, blow ratios, die swell, cooling, and blowing techniques.

Raw material companies such as DuPont, Union Carbide, Phillips Chemical, and Dow as well as universities were studying the new polymers, and test methods for melt flow, shear rate, molecular weight, tensile strength, elongation, and crystallization, etc. were being given to molders to understand their process better.

From a mere three blow molders listed in 1957, there were 97 in 1959 and 24 suppliers of blow molding machines mainly due to the growth of HDPE as a blow molding material. In 1963 it was reported to SPI that over 195 million bottles were produced (Tables 1.3 and 1.4).

In 1963, approximately three companies were supplying bottles less than eight ounces to the plastic blow molding industry, Wheaton, Brockway, and Owens-Illinois.

High density polyethylene became the dominant blow molding material in the late 1960s and is the largest volume material used in the blow molding industry today, more than 57% of the total blow molding market. PET is growing and

Table 1.2

Low Density Polyethylene Period: 1945–1958

Major Developments

1945   Low density polyethylene available in commercial quantities

1947   Owens-Illinois sells injection blow molded polystyrene bottles

1949   Mills operating horizontal wheel blow molding machine—continuous tube extrusion—8 oz to 40 oz squeeze bottles

1950   Mills patent issued for wheel machine

1950   Kautex offers first commercial blow molding machine for sale

1951   Pirelli in Italy makes carboys—extruder feeds into vertical injection cylinder

Early 1950s   Bekum, Kautex, and Fischer blow molding machines mold PVC in Europe

1951   Jackson and Church developed proplasticizer at right angle to injection machine to increase plasticizing capacity

1952   Plax manufacturing 13 gallon carboys, Hedwin 15 gallon, and Delaware Drum 100 gallon

1955   Kautex shipping blow molders to USA

1956   Reciprocating screw injection machine in use

Other Developments Important to Blow Molding

Largest injection machine in USA in 1940, 40 oz. shot and 750 ton clamp

Beryllium copper being used in mold manufacture

Machine developed for blow molding which fed individual heads by directing flow from an extruder to each head using a valving system

Horizontal and vertical wheel machines in use

Extrusion blow and injection blow in use with many versions designed primarily by the manufacturer of blow molded articles

Hydraulics improved with better valving, better hoses, improved control systems, better hydraulic motors, and less problems

Temperature controllers on-off but made more reliable

Great improvement in limit switches, timers, thermocouples, electric motors, and systems developed to control sequential motions

Only ram injection available but major changes in machine size, heating of material and molds, and controls

Two-step injection in use

Hot runner molding being practiced

Major improvements in mold making and metals

Parison programming being investigated

Stress cracking recognized as a major problem

Table 1.2 Continued

New Plastic Materials Available
  Polyethylene-isobutylene blends (1948), LDPE (1945), ABS (1948),
    SAF (1948), polyacetals (1956), high density polyethylene
    announced (1955), polyurethane (1954).
New Blow Molded Products
  Stopette bottle (1945), Squeeze bottle high volume (1950), Catsup
    dispenser (1947), drum liners, jerry cans, various type bottles

---

will continue to grow from its 33% share in 1977. Other resins, such as AN, PVC, PS, PC, LDPE, and coextrusions make up the remaining 10% of the blow molding market.

The largest single item produced from HDPE by the blow molding industry is the one-gallon milk container. Originally, custom blow molders, such as Continental Can, Owens-Illinois, Sewell Plastics, and Berwick plus others supplied the dairy industry. However, today there are more than 600 machines in operation as in-house producers, and the Uniloy blow molding machine is dominant. As it exists now the one-half gallon market is approximately 40% plastic and 60% plastic-coated paper. Within the next 5 years the one-half gallon market should be more than 75% plastic (Fig. 1.3).

In 1977, the first PET 64 oz. stretch blow molded soft drink bottle appeared in the marketplace. It was produced by Amoco in Norcross, Georgia on an RHBV produced by Cincinnati Milacron, Cincinnati, OH and was for Pepsi Cola.

The introduction of the stretch blow molded PET soda container was the most significant development since the introduction of the two-piece can. Messrs. Nathaniel Convers Wyeth and Ronald Newman Roseveare of E.I. DuPont de Nemours and Company were granted the patent for the PET stretch blow molded container on May 15, 1973 (see Appendices 1 and 2).

The PET bottle industry has grown from zero in 1977 to over 10 billion containers produced in the United States in

Table 1.3

Early High Density Polyethylene Period: 1957–1964

Major Developments

1958	First electronic parison programmer
1958	First time blow molding machine above at NPE
1958	Rigid PVC bottles used in Europe for liquid aromatics
1959	HDPE bottles for detergent, bleach, and acid cleaners accepted
1959	Calibrated neck patent
1961	Flow equalization successfully used
1961	Kautex coextrudes LDPE inside and PVC outside; adhesion poor
1962	Carbon dioxide cooling patent
1963	Two-step blow molding in Europe (extrude parisons, blow later)
1963	Double-wall containers marketed
1963	First patent for coextrusion and blow molding
1964	Very large wheel machine for one-gallon bottles operated
1964	First milk in-house bottle manufacturing
1964	Marrick process in England for PVC

Other Developments Important To Blow Molding

Very good injection blow machines developed; many commercial extrusion presses blow bottle machines in production; large number of types of extrusion blow molding machines offered for sale to blow molders; large blow molding machines in production; milk and lubricating oil packaging major targets for blow molding.

Blow molding machines using vertical and horizontal reciprocating screw injection, twin screw, and single screw extruders, many types of screws, accumulators, manifolds, rotary molding system, wheels, shuttle systems, rising molds, descending molds, blow air controls, parison programming, neck finishing operations, blown part handling systems, bulk material handling systems, flame treating, leak detectors, variable orifices, shaped orifices, extrusion rate programming, calibration necks, handle ware, surface treatments with flame, corona discharge, and chemicals for decorating with many new techniques, tube cutoff methods-knife, hot wire, and flame.

Molds made from steel, beryllium-copper, aluminum, Kirksite, and cast iron. Venting and cooling improved. Various surface treatments and coatings tried experimentally to improve barrier or other properties.

Table 1.3 Continued

New Plastic Materials Available
   High density polyethylene (1958), polycarbonate (1957), phenoxy
      (1962), polypropylene (1958), polyallomer (1962), ionomer
      (1964), polypropylene oxide (1964), ethylene-vinylacetate (1964).
New Blow Molded Products
   HDPE nurser, handled bottles, aerosols, double-wall bottles, Breck
      Shampoo (1958), PVC in Europe for liquid aromatic bottles
      (1958), bottles for noncarbonated vegetable and fruit juices, PVC
      for solvent-containing products, plumbing products, toilet floats,
      lawn mower rollers, ducting, small tanks, bellows, textile hobbies,
      seats, containers for broad range of food products, large
      moldings, toys, building products, refrigerators, automotive,
      airplanes, drums

1997. The PET container market is forecast to grow 5%–7%
for the next five years. The remaining large markets for PET
are beer, baby food, sauces, pickles and fruits (glass).

   In the early stages of the PET stretch blow molded container market, a new PET container line could be installed

Table 1.4

Major Developments 1955–1970

1965	Polyacetal bottles used for aerosols
1968	Orbet process for stretch blow molding
1968	Polypropylene used for detergents
1968	FVC bottles marketed for liquor but unacceptable to FDA
1969	First testing acrylonitrile type plastics for carbonated beverages
1970	Acrylonitrile type plastic used for sundae topping, salad oil, syrups, vinegar, and wines
1970	Aseptic containers used
1970	Parison programmers used on many blow molding machines
1970	Orbet process
1970	Coextrusion actively investigated in USA

New Plastic Materials Available
   Thermoplastic polyester (polyethylene teraphthalate) (PET) (1970),
      polysulfone (1965), modified polyacrylonitriles.

Figure 1.3 Uniloy machine producing half-gallon HDPE milk containers, swing arm assembly. The bottle is taken away to the cooling conveyor 8 head. (Courtesy of Johnson Controls, Inc., Plastics Machinery Division, Manchester, MI.)

for under one-half million dollars utilizing the single-stage Nissei (Japanese) machine. Today, a similar line costs close to $1 million. The high volume markets have predominantly all switched from single-stage to two-stage production production. In the single-stage machines, all of the production, injection molding, conditioning, stretch blow molding, and discharging is done in one machine. In the two-stage process the preform is injection molded on an injection machine and cooled to room temperature and either used later in the same facility or packed and shipped to outlying locations where the reheat stretch blow molding machines are located. Reheat stretch blow molding machines produce from 800 containers per day (8 hours) up to 42,000 containers per hour. The Sidel/Husky machines are used to produce over 65% of all reheat stretch blow molded containers in the world market.

There has really been no significant blow molding development since 1977 in the PET market. Heat setting of PET for packaging isotonic drinks has been developed and is special as are blow molded angioplasty balloons (see Tables 1.5–1.7).

The largest growing blow molding market at present is the coextruded blow molded gasoline tanks for use in the automotive industry. It is forecast by the year 2002 that 80% of U.S. manufactured automobiles will feature six layer tanks (HDPE with adhesives and EVOH-ethylene vinyl alcohol) The second largest growing market is the 55-gallon drum market whether it be monolayer or multilayer. (Figs. 1.4 and 1.5).

1.3 CONCLUSION

Naturally, the blow molding machines will get larger. They will continue to upgrade their user-friendly programmable controls and don't be surprised to see an all electric blow molding machine within the next two to three years. We are due for a major advance.

Table 1.5
1971–1978

Major Developments

1972   Toya Selkan develops multilayer blow molding process
1973   PET bottle patented
1973   Inflatable rubber balloon process
1973   First methodology for surface cooling blow molded parts
1974   Biaxially oriented processes become major investigation
1974   Solid-state controls replacing other controllers
1975   Two-step commercial biaxial orienting machines available
1976   First Pepsi: PET bottle production
1977   Japanese offer one-step biaxially orienting machines
1977   55-gallon drum in volume production
1977   Two-step injection blow molding with rotation (MWR)

Other Developments Important to Blow Molding

Production equipment for the Orbet process (Bi-Axomat); four station injection blow rotaries; For PVC, injection blow shorter plasticizing screw, lower compression ratio screw, streamlined nozzle, and special manifolds; improved controllers for temperature, timing, pressures, and all aspects of a blow molding machine and more solid state; accumulator head improvements and annular pistons used; all hydraulic operations; major cooling improvements; in-mold labeling being used; larger and larger blow moldings.

High volume PET stretch blow machines in use; increased activity in multilayer; greater use of other plastics, including engineering plastics; HXW-FE powder used for large moldings; used machinery business increasing rapidly; displacement blow developed; auxiliary equipment for blow molding becoming extremely important.

New Plastic Materials Available

Lopac (1975), cryolite (acrylic multipolymer) (1977), PVC-ethylene vinyl chloride copolymer, high clarity PVC (1978), cycopak (ABS based high nitrile resin), nitrile barrier resins (1975), polybutylene (1975), polyarylate (1975), styrenic terpolymers (1972), HXV-PE, acrylic modified nylon 6

New Blow Molded Products

Carbonated beverage bottles, coextruded fuel tanks, water tanks, trash containers, furniture, household articles, recreational items, objects for agriculture, cooler chests, polycarbonate nursers, large toys, ductwork

Table 1.6

Most Outstanding Developments: 1930–1991

Technology and equipment	Molding materials
Use of the wheel	Low density polyethylene
Calibrating necks	High density polyethylene
Programming parisons	Polypropylene
Stretch blow molding (orientation)	Polyethylene terephthalate
Orifice shape control	Polycarbonate
Head tooling	Improved polyvinyl chloride
Injection rotaries	Ethylene vinyl acetate
Computerization	Copolymers
System for blowing heat sensitive	Alloys
materials	Control of molecular weight and
Science of polymer technology	molecular weight distribution
Pressblow	
In-mold labeling	
Reciprocating screw	
Multilayer	
Testing equipment	
Coextrusion	
Coinjection	
Accumulators	
Manifolds	
Study of rheology	

Table 1.7

1977 to the Present

1977 PET 64 oz. container
1981 Three layer coextrusion—"gamma"
1984 Hot fill PET (heat set)
1984 Quart oil containers
1984 Five-gallon PC water container
1985 PET liquor containers
1985 PET wide mouth
1986 Multilayer plastic cans
1988 First production heat set PET
1992 Coinjection
1992 Plastic automotive gas tanks

New Products

Ketchup containers in plastic (gamma then PET), PC light globes, automotive bellows (TPEs), automotive spoilers (PC and Bexloy), toys (Rubbermaid and Little Takes) from HDPE, angioplasty balloons (PET), drums (HDPE), gasoline tanks (HDPE coextruded), garbage cans (HDPE), furniture (HDPE).

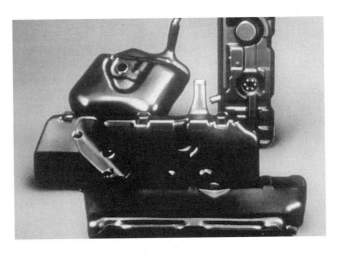

Figure 1.4 Multilayer gas tank.

Figure 1.5 Monolayer drum.

APPENDIX 1

Detroit section SPE charter members and newsletter (Courtesy of R.E. Dunham).

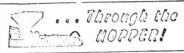

Items for THROUGH THE HOPPER! are welcome.

Joe Hutt has joined DME in charge of publications. The best of everything to you, Joe.

Carl Whitlock has a new Florida address if you want to write or stop in. FF-32, Channel Cay Drive, Key Largo, Fla. 33037.

Amaco Chemical Corp., Industrial Products Division, Stow, Ohio, is looking for a process engineer with PVC extrusion, blending and formulation background. If interested, call A.C. Beckwith, 216 / 688-8291.

Darrel Pufahl now with AMACO Detroit, replacing Dick Stae.

Merv Crozier has joined Don Lico at Fisher Body as his assistant, good luck Merv.

Plumbing Heating Division, American Standard, New Brunswick New Jersey looking for chemical engineer experienced in thermosetting. $18,000—20,000 salary. Call Gordon McKiernan, 201 / 885-1900 for information.

Ed Greene, Design Models, has his back room all closed off to anyone except a list of employees and the U.S. Tank Center. Shades of WW II.

Last issue for the year — thanks for all your help.

George Thom

SPE PLANT TOUR — APRIL 2

THIS IS WHAT 138 DETROIT AND MID-MICHIGAN SPE MEMBERS AND GUESTS FOUND.

Baker Plastics, a division of Haskon, Inc., wholly owned subsidiary of Hercules Inc., manufactures Blowmolding Equipment and produces Blowmolded Containers as an adjunct to its equipment business. Over 125 hourly, and 25 salaried personnel are employed in our Williamston facility, which includes (with our new additions) 120,000 square feet of manufacturing and office space on a 50 acre tract of land. Baker Plastics Blowmolding Systems are sold throughout the United States, to both custom blow molders and to companies desiring in-plant manufacturing capabilities.

Baker Plastics uses the total systems approach in determining customers' requirements. Material storage and handling systems, chilling equipment, mechanical and electrical equipment requirements, inline container storage systems and fully automatic equipment for molding and finishing of blow-molded parts, are available from the Baker Plastics Organization.

Economic analysis of projected unit costs and return on investment, engineering for single machine lines or a complete operational plant, design and qualification of parts, installation of all equipment and subsequent service, are offered by the baker Plastics Organization.

Baker Plastics Equipment features the FLOATING PLATEN CLAMP, a concept that permits the continuous extrusion of high quality parisons (either round or with extreme ovalization). Molds are transferred laterally with the captured parison to the molding stations. HYDRAULIC DRIVE results in a continuous extrusion over an infinitely variable range of screw speeds at minimum noise levels. Baker Plastics' unique PREFINISHING SYSTEM, with VACUUM ASSIST, results in quality neck prefinishes. SOLID STATE PROGRAMMING and CO2 SYSTEMS are available as optional equipment to provide optimum cycle time and lowest product cost.

APPENDIX 2

Sample patent.

UNITED STATES PATENT AND TRADEMARK OFFICE

Certificate

Patent No. 3.733,309   Patented May 15, 1973

Nathaniel Convers Wyeth, and Ronald Newman Roseveare

Application having been made by Nathaniel Convers Wyeth, and Ronald Newman Roseveare, the inventors named in the patent above identified, and E. I. du Pont de Nemours and Company, Wilmington, Del., the assignee, for the issuance of a certificate under the provisions of Title 35, Section 256, of the United States Code, adding the name of Frank P. Gay as a joint inventor, and a showing and proof of facts satisfying the requirements of the said section having been submitted, it is this 11th day of March 1975, certified that the name of the said Frank P. Gay is hereby added to the said patent as a joint inventor with Nathaniel Convers Wyeth, and Ronald Newman Roseveare.

Associate Solicitor

PO-1050
(5/69)

UNITED STATES PATENT OFFICE
CERTIFICATE OF CORRECTION

Patent No. __3,733,309_____   Dated __May 15, 1973_____

Inventor(s)__Nathaniel Convers Wyeth and Ronald Newman Roseveare__

It is certified that error appears in the above-identified patent and that said Letters Patent are hereby corrected as shown below:

Column 1, line 66, "thepo lymer" should read -- the polymer --.
Column 2, line 55, "anuular" should read -- annular --.
Column 3, line 66, "catalyst" should read -- catalysts --.
Column 5, lines 8 and 9, "cabonation and water in the beverage yet keep ou" should read -- carbonation and water in the beverage, yet keep out --.
Column 6, line 60, "behaivor" should read -- behavior --.
         line 65, "srain" should read -- strain --.
Column 8, line 32, "bifuracted" should read -- bifurcated --.
Column 10, line 23, "FIG. 3" should read -- FIGS. 3 and 4 --.
Column 11, line 72, "to" should read -- into --.
Column 13, line 45, "where in" should read -- wherein --.
Column 14, line 21, "extrusions" should read -- extrusion, --.
         line 39, "slug first" should read -- slug is first -
         line 70, "ordet" should read -- order --.
Column 15, line 21, "polyproplylene" should read
-- polypropylene --.
Column 16, line 17, the entire line is misplaced and should be the first line appearing after "Example 3".
Column 16, line 49, "2 Ø peak" should read -- 2 θ peak --.
         line 51 (third line in Table III), "46 (hoop)" should read -- 66 (hoop --.
Column 16, line 53, "hopp" should read -- hoop --.
         line 54 (last line in Table III), "87°X" should read -- 87°Ø --.
Column 16, line 68, "p.s. ig." should read -- p.s.i.g. --.
         line 74, "sacrified" should read -- sacrificed --.
Column 17, line 5, "scopet hereof" should read -- scope thereof --.
Column 17, line 6, "embodimens" should read -- embodiments --.
         line 8, "equavilence" should read -- equivalence --
         line 13 (Claim 1), "1,331 to 1,402" should read -- 1.331 to 1.402 --.
Column 18, line 20 (Claim 9), immediately preceding "5%" the phrase -- autogenous pressure of about 75 p.s.i.g. is no greater than -- is missing.

Signed and sealed this 4th day of December 1973.

Rene D. Tegtmeyer

RENE D. TEGTMEYER
Acting Commissioner of Patents

ATTEST:

Edward . . .

ATTESTING OFFICER

May 15, 1973   N. C. WYETH ET AL   **3,733,309**

BIAXIALLY ORIENTED POLY(ETHYLENE TEREPHTHALATE) BOTTLE

Filed Nov. 30, 1970   6 Sheets-Sheet 1

F I G. 1

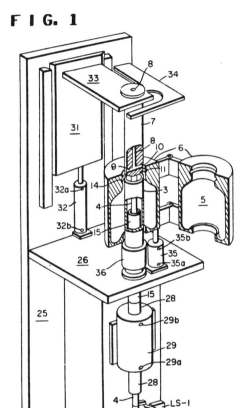

INVENTORS
NATHANIEL CONVERS WYETH
RONALD NEWMAN ROSEVEARE

BY Francis A. Painter

ATTORNEY

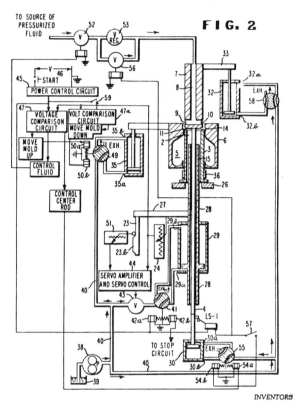

May 15, 1973   N. C. WYETH ET AL   3,733,309
          BIAXIALLY ORIENTED POLY(ETHYLENE TEREPHTHALATE) BOTTLE
Filed Nov. 30, 1970   6 Sheets-Sheet 2

FIG. 2

                                                                        INVENTORS
                                                    NATHANIEL CONVERS WYETH
                                                    RONALD NEWMAN ROSEVEARE

                                                    BY Francis A. Painter
                                                                        ATTORNEY

May 15, 1973   N. C. WYETH ET AL   **3,733,309**

BIAXIALLY ORIENTED POLY(ETHY ENE TEREPHTHALATE) BOTTLE

Filed Nov. 30, 1970   6 Sheets-Sheet 3

FIG. 3   FIG. 4

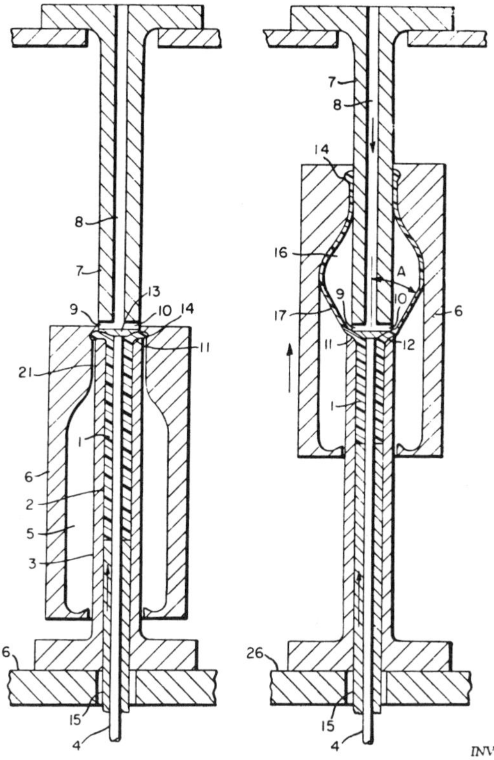

INVENTORS
NATHANIEL CONVERS WYETH
RONALD NEWMAN ROSEVEARE

BY *Francis A. Painter*

ATTORNEY

May 15, 1973 N. C. WYETH ET AL **3,733,309**

BIAXIALLY ORIENTED POLY(ETHYLENE TEREPHTHALATE) BOTTLE

Filed Nov. 30, 1970 6 Sheets-Sheet 4

FIG. 5

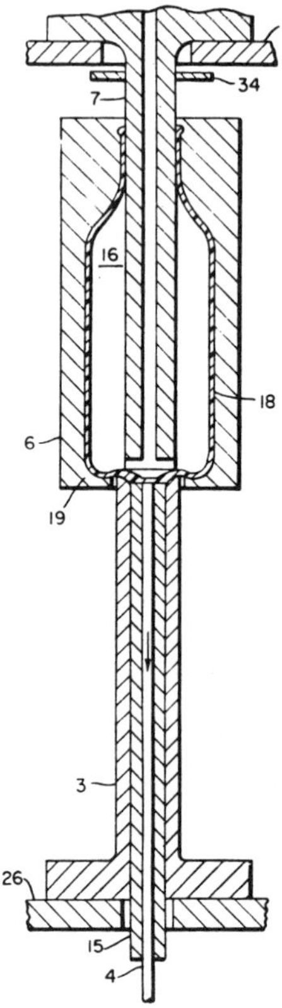

FIG. 6

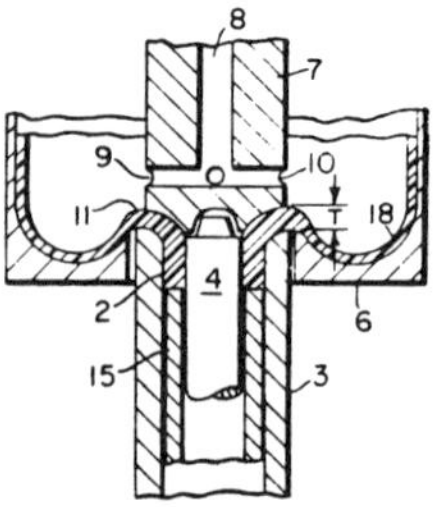

FIG. 7

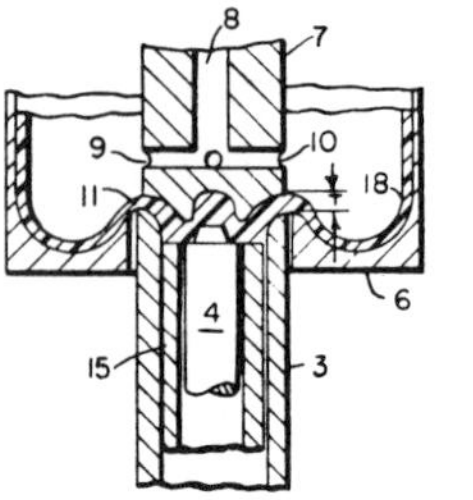

INVENTORS
NATHANIEL CONVERS WYETH
RONALD NEWMAN ROSEVEARE

BY Francis A. Paintin
ATTORNEY

May 15, 1973   N. C. WYETH ET AL   **3,733,309**

BIAXIALLY ORIENTED POLY(ETHYLENE TEREPHTHALATE) BOTTLE

Filed Nov. 30, 1970   6 Sheets–Sheet 5

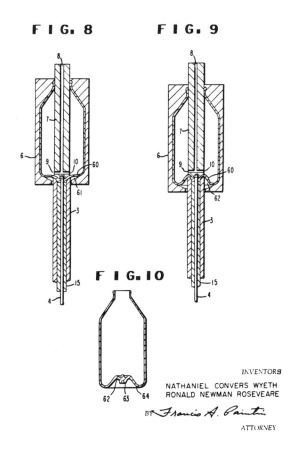

FIG. 8   FIG. 9

FIG. 10

INVENTORS
NATHANIEL CONVERS WYETH
RONALD NEWMAN ROSEVEARE

BY Francis A. Painter

ATTORNEY

May 15, 1973   N. C. WYETH ET AL   3,733,309
BIAXIALLY ORIENTED POLY(ETHYLENE TEREPHTHALATE) BOTTLE
Filed Nov. 30, 1970   6 Sheets—Sheet 6

F I G. 11   F I G. 12   F I G. 13

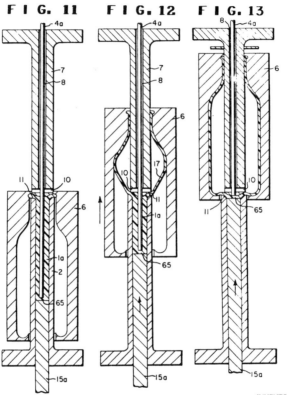

INVENTORS
NATHANIEL CONVERS WYETH
RONALD NEWMAN ROSEVEARE

Francis H. Vaintin
ATTORNEY

United States Patent Office

3,733,309

Patented May 15, 1973

1

3,733,309
**BIAXIALLY ORIENTED POLY(ETHYLENE
TEREPHTHALATE) BOTTLE**
Nathaniel Convers Wyeth, Mendenhall, Pa., and Ronald
Newman Roseveare, Wilmington, Del., assignors to
E. I. du Pont de Nemours and Company, Wilmington,
Del.
Continuation-in-part of abandoned application Ser. No.
885,853, Dec. 17, 1969. This application Nov. 30, 1970,
Ser. No. 93,571
Int. Cl. B29c 5/06
U.S. Cl. 260—75 T   14 Claims

ABSTRACT OF THE DISCLOSURE

A hollow, biaxially oriented, thermoplastic article, particularly a bottle, prepared from polyethylene terephthlate wherein said article has an inherent viscosity of at least 0.55, a density of about 1.331 to 1.402 and a ratio of article weight in grams to volume in cubic centimeters of about 0.2 to 0.005:1. Such articles have excellent strength properties, are impact resistant, and are capable of holding liquids under pressures as high as about 100 p.s.i.g. at a temperature of about 50° C. without significant deformation making such articles useful in bottling liquids under pressure such as beer, carbonated beverages, or aerosols.

CROSS REFERENCE TO RELATED CASE

This application is a continuation-in-part of U.S. patent application Ser. No. 885,853, filed Dec. 17, 1969 and now abandoned.

BACKGROUND OF THE INVENTION

This invention relates to a hollow, biaxially oriented thermoplastic article, such as a bottle, particularly useful in bottling liquids under pressure, i.e. carbonated beverages or aerosols, etc., along with a method and apparatus for its production.

Thermoplastic bottles useful in bottling carbonated beverages wherein the bottle contains some autogenous pressure, and must have the properties necessary for such use. These properties include the necessary strength requirements to contain the beverage under pressures which can be as high as 100 p.s.i.g. without appreciable creep or serious deformation within the in-use temperature range of about 0 to 50° C. In addition, the bottle must have a low permeability rating, particularly with respect to carbon dioxide and oxygen. The continuous loss of carbon dioxide from a carbonated beverage, or the infusion of oxygen into a beverage such as beer, shortens the shelf life and alters the taste of the beverage.

Several methods are known for forming hollow thermoplastic articles. One method is blow molding. In conventional blow molding, a parison is first formed by extruding a heat-softened thermoplastic tube and pinching off the bottom or, alternatively, by injection molding of the blowable geometric form directly. Secondly, the parison or blowable form is then contained within a mold cavity having the volumetric configuration of the desired thermoplastic article and the parison is expanded by blowing it with compressed air within the confines of the mold cavity. The polymer is extruded and blown at elevated temperatures, namely above the orientation temperature range of thepo lymer. The article produced is not biaxially oriented. This process is economical but it is relatively slow and, more importantly, the hollow articles produced lack the necessary strength for use in bottling beverages under pressure.

Other prior art methods involve the use of solid plastic slugs which are extruded to form a hollow article. When

2

a solid slug is extruded, the material that was originally situated at the slug axis of symmetry ultimately appears on the inner wall of the article being made and exhibits extreme roughness and other surface defects. The use of higher temperatures during the article formation tends to alleviate these adverse features but use of a high temperature is self-defeating, since possible strength gained from biaxial orientation of the plastic material during bottle formation is lost by heat relaxation.

These is a need, therefore, for an economical process with the necessary apparatus for producing a plastic bottle having characteristics which make it useful in bottling liquids under pressure such as carbonated beverages or aerosols.

SUMMARY OF THE INVENTION

Accordingly, the present invention provides a method and apparatus for producing a hollow, biaxially oriented, thermoplastic article having improved strength properties. The articles are biaxially oriented by stretching typically an average of up to about 4.0 times in the axial direction and about 2.5 to 7.0 times in the hoop direction. The preferred thermoplastic employed is polyethylene terepthalate having an inherent viscosity of at least 0.55. The article formed not only has improved strength properties, but also has reduced permeabilities to carbon dioxide, oxygen and water, making the process and apparatus particularly suitable for producing thermoplastic bottles useful for bottling beverages under pressure, such as carbonated sodas and beer. Polyethylene terephthalate bottles produced according to this invention have a density within the range of about 1.331 to 1.402 and the right cylinder section of the bottle has an axial tensile strength of about 5,000 to 30,000 p.s.i.; a hoop tensile strength of about 20,000 to 80,000 p.s.i.; an axial yield stress of at least 4,000 p.s.i.; and a hoop yield stress of at least 7,000 p.s.i. Typically, these bottles will have a shell thickness of about 10 to 30 mils, a ratio of weight in grams to volume in cubic centimeters of about 0.2 to 0.005:1, and a deformation constant equal to the slope of the log (reciprocal of the strain rate) versus strain having a value of at least about 0.65.

DESCRIPTION OF THE DRAWINGS

FIG. 1 is a perspective view of the apparatus of this invention, including means for actuating moving parts of the apparatus.

FIG. 2 is a schematic diagram of the principal parts of the apparatus of FIG. 1 showing hydraulic, fluidic and electrical circuits for actuating and controlling the apparatus.

FIG. 3 is a fragmentary cross-sectional view of the apparatus of this invention positioned during the initial stage of forming a hollow article and specifically showing an annuular bead being formed.

FIG. 4 is a fragmentary cross-sectional view of the apparatus of this invention positioned during the intermediate stage of forming a hollow article and specifically showing the crucial step of combined nonmolten extrusion and expansion by use of internal fluid forces.

FIG. 5 is a fragmentary cross-sectional view of the apparatus of this invention positioned after the hollow article has been completely formed.

FIG. 6 is an enlarged fragmentary cross-sectional view of a portion of the positioned apparatus shown in FIG. 5 showing in greater detail the region around the annular extrusion orifice near the completion of the combined extrusion and expansion operation.

FIG. 7 is an enlarged fragmentary cross-sectional view similar to that of FIG. 6 but showing the region around the annular extrusion orifice upon the completion of the hollow article formation.

3,733,309

3

FIGS. 8 and 9 are fragmentary cross-sectional views of the apparatus of the invention adapted to handle polymers having poor conformability to the mold.

FIG. 10 is a bottle formed in the apparatus of FIG. 9.

FIG. 11 is a fragmentary cross-sectional view of an alternate embodiment incorporating a sliding center rod with the center rod in its fully extended position.

FIG. 12 is a fragmentary cross-sectional view of the alternate embodiment shown in FIG. 11 with the center rod at an intermediate stage of withdrawal corresponding relatively to the intermediate position of the sliding mold.

FIG. 13 is a fragmentary cross-sectional view of the alternate embodiment shown in FIG. 11 with the center rod completely withdrawn corresponding relatively to the final position of the sliding mold.

DETAILS OF THE INVENTION

Polyethylene terephthalate articles of this invention are generally cylindrical, typically in the shape of a soda bottle or a beer bottle, biaxially oriented, having densities ranging from about 1.331 to 1.402, and can be made to be transparent and glossy with no haze, or they can be dyed by the addition of a dye to the polymer. In addition, the articles have physical properties which make them very useful for containing liquids under pressure. These physical properties include high tensile strength, low creep at low weight to volume ratios, high resistance to impact (toughness), and good permeation qualities.

Polyethylene terephthalate useful in preparing the thermoplastic articles of this invention includes (a) polymers wherein at least about 97% of the polymer contains the repeating ethylene terephthalate units of the formula:

$$-OCH_2CH_2O-\overset{O}{\underset{\parallel}{C}}-\overset{}{\underset{}{\bigcirc}}-\overset{O}{\underset{\parallel}{C}}-$$

with the remainder being minor amounts of ester-forming components, and (b) copolymers of ethylene terephthalate wherein up to about 10 mole percent of the copolymer is prepared from the monomer units of diethylene glycol; propane-1,3-diol; butane-1,4-diol; polytetramethylene glycol; polyethylene glycol; polypropylene glycol; 1,4-hydroxymethylcyclohexane and the like, substituted for the glycol moiety in the preparation of the copolymer or isophthalic; bibenzoic; naphthalene 1,4- or 2,6-dicarboxylic; adipic; sebacic; decane-1,10-dicarboxylic acid, and the like, substituted for the acid moiety in the preparation of the copolymer.

The specific limits on the comonomer are governed by the glass transition temperature of the polymer. It has been found that when the glass transition temperature extends below about 50° C., a copolymer having reduced mechanical properties results. Accordingly, this corresponds to the addition of no more than about 10 mole percent of a comonomer. One exception to this, for example, is the addition of bibenzoic acid where the glass transition temperature of the copolymer remains above 50° C. and does not drop with the addition of more than 10 mole percent. Others would be obvious to those skilled in the art.

In addition, the polyethylene terephthalate polymer can include various additives that do not adversely affect the polymer in use such as stabilizers, e.g., antioxidants or ultraviolet light screening agents, extrusion aids, additives designed to make the polymer more degradable or combustible, such as oxidation catalyst, as well as dyes or pigments.

The polyethylene terephthalate should have an inherent viscosity (1% concentration of polymer in a 37.5/62.5 weight percent solution of tetrachloroethane/phenol, respectively, at 30° C.) of at least 0.55 to obtain the desired end properties in the articles formed and preferably the inherent viscosity is at least about 0.7 to obtain an article having excellent toughness properties, i.e., resistance to impact loading. The viscosity of the polymer

4

solution is measured relative to that of the solvent alone and the

Inherent viscosity

$$= \frac{\text{natural logarithm} \dfrac{\text{viscosity of solution}}{\text{viscosity of solvent}}}{C}$$

where C is the concentration expressed in grams of polymer per 100 milliliters of solution.

Biaxial orientation of the articles of the present invention is useful to impart improved physical properties such as improved tensile strength and yield strength. Biaxial orientation is accomplished by stretching the thermoplastic in the axial and hoop directions as the article is being formed. The article of the present invention is molecularly oriented by being biaxially stretched an average of about up to 4.0 times in the axial direction and about 2.5 to 7.0 times in the hoop direction. Preferably, such stretching is carried out at the orientation temperature of the thermoplastic, i.e. above the glass transition temperature and below the crystal melting point. The extent of the molecular orientation can be determined by known techniques. One method of determining molecular orientation is described in The Journal of Polymer Science, vol. XLVII, pages 289–306 (1960), entitled "X-Ray Determination of the Crystallite Orientation Distribution of Polyethylene Terephthalate Films," by C. J. Heffelfinger and R. L. Burton; and "Structure And Properties of Oriented Poly(ethylene Terephthalate) Films," by Heffelfinger and Schmidt in the Journal of Applied Polymer Science, vol. 9, page 2661 (1965). Both articles are hereby incorporated by reference.

Biaxial orientation is instrumental in providing excellent strength properties. Articles prepared according to the present invention typically do not have the same degree of orientation at every point on the article; however, the areas that are less oriented have a thicker shell than the areas that are more highly oriented, thereby providing relatively high overall strength to the article. In preparing a bottle, the thinnest shell thickness occurs in the right cylinder section; however, this section is the most highly oriented. In the right cylinder section of a bottle prepared by stretching, in accordance with this invention, the tensile strength and yield stresses typically are: axial tensile strength of about 5,000 to 30,000 p.s.i.; a hoop tensile strength of about 20,000 to 80,000 p.s.i.; an axial yield stress of at least 4,000 p.s.i. and a hoop yield stress of at least 7,000 p.s.i. The values of tensile strength and yield stress were determined by the method described in ASTM D–882, entitled "Tensile Testing."

The density (grams per cubic centimeter) of the article can range from about 1.331 to 1.402. Density was measured by the method described in ASTM 1505, entitled "Density Gradient Technique." Density is a measure of the crystallinity and this density range includes a crystallinity range of about 0 to 60%. The percent crystallinity is calculated from the following equation:

$$\text{Percent Crystallinity} = \frac{Ps - Pa}{Pc - Pa} \times 100$$

where

Ps = density of test sample (g./cm.^{3})

Pa = 1.333 (g./cm.^{3}), density of amorphous film of zero percent crystallinity

Pc = 1.455 (g./cm.^{3}), density of the crystal calculated from unit cell parameters.

The formed articles themselves can have varying crystallinities along the axial length of each article, in which case, if desired, the article can be heat-set to achieve a uniform crystallinity in each article.

Orientation and crystallinity each contribute to certain properties; however, under some conditions, they are competitive. For example, increased orientation provides increased tensile properties but tends to decrease the

3,733,809

5

thermal stability of the article. To offset the latter, the bottle can be heat set to increase the crystalliniy of the bottle.

Crystallinity is also related to the barrier properties of the article, particularly permeation properties. In bottling carbonated beverages under pressure, such as soda or beer, for example, it is important that the bottle have barrier properties sufficient to contain the cabonation and water in the beverage yet keep ou contaminants such as oxygen.

It has been found that increasing the crystallinity lowers the ability of carbon dioxide, oxygen or water vapor to permeate the bottle. The term "permeate" and its derivatives, used in this application, mean the ability of an agent such as carbon dioxide, oxygen or water vapor to pass through or diffuse through the shell of the articles of this invention. The degree of permeation encountered during the use of a bottle is dependent on many variables including the total surface area of the bottle, ambient temperature, the pressure inside the bottle, and the type and amount of liquid in the bottle.

When the crystallinity of the bottle is at least about 15% (density approximately 1.348), and the bottle is being used in a conventional manner for bottling sodas or beer in the usual consumer size bottle, i.e. 6-, 8-, 10-, 12-, or 16-ounce bottles, the degree of permeation related to the various permeating agents of concern is sufficient to meet commercial standards. For example, in bottles containing up to about 16 ounces of soda or beer under about 75 p.s.i.g. autogenous pressure at room temperature, i.e. about 25° C., wherein the shell thickness is between 10 and 30 mils, and a ratio of weight in grams to volume in cubic centimeters of about 0.2 to 0.005:1, the carbon dioxide leaving the bottle is no more than 15% in 30 days, the oxygen permeation through the shell into the liquid is no greater than 5 parts per million in 30 days and the amount of water lost from the liquid is no greater than 5% in 90 days.

The carbon dioxide permeation is measured by pressurizing a bottle with 75 p.s.i.g. of carbon dioxide, capping the bottle by conventional capping means, placing the pressurized bottle in a vacuum chamber where the vacuum is one micron of mercury, allowing the bottle in the vacuum chamber to equilibrate, then measuring the pressure rise in vacuum chamber as a function of time. Alternatively, the same pressurized bottle can be placed in a closed chamber with a stream of nitrogen made to pass by the bottle. Thereafter the stream of gas is scrubbed in a sodium hydroxide bath and titration of standard sodium hydroxide will indicate the amount of carbon dioxide picked up by the passing stream of nitrogen. The amount of carbon dioxide measured per unit of time provides the rate of carbon dioxide permeation.

The oxygen permeation is measured by filling a bottle with degassed water, sealing the bottle by conventional means, storing the bottle at room temperature and pressure and periodically measuring the oxygen content of the water inside the bottle by a known technique, i.e. silver electrode potentiometric titration.

The water permeation is measured by placing a dessicant inside of a dry bottle, sealing the bottle, storing the bottle at 37.8° C. in an atmosphere having a constant relative humidity of 100% then periodically weighing the bottle to determine the amount of water picked up by the dessicant. Alternatively, the bottle can be filled with water, pressurized to an autogenous pressure of 75 p.s.i.g. and capped, then placed in an atmosphere having a relative humidity of about 15% at 25° C. and periodically weighed for water loss.

Another significant property important to acceptability of the articles of this invention for use in bottling liquids under pressure, is that the bottle exhibits relatively low creep, particularly in thin walled, low weight articles. Creep is the change in structural dimension of the article on exposure to stress and is related to many factors

6

including the stress level, the type of polymer, the physical state of the polymer, the ambient temperature and the time of exposure to stress. When considering creep in a generally cylindrical bottle, the size and shape of the bottle is also significant. In addition, the autogenous pressure in the bottle increases with increased temperature so creep resistance must be relatively constant over a reasonable use range of temperature and pressure. For typical uses such as bottling beer or soda, this temperature range is about 0 to 50° C. and the pressure range is about 0 to 100 p.s.i.g.

The stress levels encountered in a bottle used to contain a pressurized liquid, such as a carbonated beverage, are directly proportional to the autogenous pressure in the bottle, the diameter of the bottle and inversely proportional to the wall thickness.

The stress can be closely approximated by the expressions:

$$\sigma_{Hoop}=Pr/t$$

$$\sigma_{Axial}=Pr/2t$$

where

σ=stress
P=autogenous pressure
r=radius of right cylinder
t=shell thickness.

Typically, a bottle having a diameter of about 2.00 inches with a right cylinder wall thickness of about 20 mils at room temperature and pressurized to about 75 p.s.i.g. will be exposed to and resist a hoop stress of about 3,750 p.s.i.

Thin shelled bottles are desirable, since this means the use of less polymer, making the bottle cheaper to manufacture. However, thin shells lead to increased stress levels and a need for greater creep resistance. Biaxially orienting a polymer, other factors remaining the same, increases the yield stresses of the bottle and is, therefore, a significant reason for orientation.

Creep is usually measured on polymers by placing a sample under a fixed load, i.e. stress, at a constant temperature and measuring the strain deformation as a function of time. The curves for thermoplastics have a characteristic shape in which the rate of strain decreases as a function of time. A plot of the log (reciprocal of the strain rate) versus strain results in a linear plot over a substantial part of the creep curve. The slope of the straight line segment herein referred to as the deformation constant, is mathematically expressed as:

$$DC=\frac{d \log (dt/d\epsilon)}{d\epsilon}$$

where

DC=deformation constant
dt=differential of time
$d\epsilon$=differential of the strain.

This deformation constant is applicable to related thermoplastics and can be used to compare the creep behavior by comparing the slope values. A deformation constant equal to 0 indicates that the sample being tested is extending at its natural strain rate or for the load indicated, the srain rate is constant. A deformation constant of infinity indicates that there is no measurable strain indicated.

For bottles prepared according to the present invention, the deformation constant is at least about 0.65 indicating a deformation of less than 5% in 100 hours at 50° C. with an autogenous pressure of 75 p.s.i.g.

Still another property of the biaxially oriented polyethylene terephthalate articles of this invention is toughness or impact resistance. However, it is particularly related to the inherent viscosity of the polyethylene terephthalate. In general, increasing the inherent viscosity in-

3,733,309

7

creased the impact resistance of the bottle. This is illustrated by a drop test wherein a bottle is filled and capped under typical bottling conditions with an autogenous pressure of 60 p.s.i.g. The bottle is then dropped on a concrete floor so that the point of impact is on the edge of the base. In testing bottles similarly prepared except for their inherent viscosity, it is found that in drops at 0° C. (a) bottles having an inherent viscosity of 0.85 on the average survive a six-foot drop but fail, i.e. crack or break open, on an eight-foot drop; (b) bottles having an inherent viscosity of 0.95 on the average survive two drops at eight feet but fail on the third; and (c) bottles having an inherent viscosity of 1.1 survive five drops at eight feet.

The apparatus useful in preparing the articles of this invention will be described in detail with the aid of the drawings. Referring to FIGS. 3 to 5, a hollow, cylindrically shaped, thermoplastic polymeric slug 1 described below is first placed in an extrusion chamber 2 formed by the bore of an extrusion barrel 3 and the outside cylindrical surface of a center supporting rod 4. A mold cavity 5 of mold 6 has an internal configuration such as the shape of the article desired and is positioned in a first location surrounding the extrusion barrel 3 as is particularly shown in FIGS. 1 and 3. The mold cavity 5 illustrated in FIG. 3 is one for use in fabricating a narrow neck bottle such as can be employed in bottling carbonated beverages.

The extrusion barrel 3 is in axial alignment with a mandrel 7 having a uniform outside diameter that is substantially the same as the inside diameter of the neck of the bottle being fabricated. A fluid passage 8 is contained within the mandrel 7 having fluid exit ports 9 and 10 at the end of the mandrel 7 that is in closest proximity to the extrusion barrel 3. Situated between the end of the barrel 3 and the end of the mandrel 7 is an annular extrusion orifice 11. This orifice can conveniently be formed by rounded end 12 of the extrusion barrel 3 and annular flared piece 13 which is attached to the body of the mandrel 7.

The annular extrusion orifice 11, shown in detail in FIGS. 6 and 7, is defined by the confronting end portions of the extrusion barrel 3 and the mandrel 7. In cross-sectional profile, both members are machined with a curving shape to provide a smooth transition from the annular extrusion chamber 2 outwardly and to provide a boundary for the annular extrusion orifice 11 such that the orifice is always convergent in its cross-sectional area. The orifice becomes progressively smaller in the direction of flow out to the outer boundary of its extrusion annulus which is proximate to the periphery of the mandrel 7 and from which the polymer emerges from the annular extrusion orifice 11 to enter the cavity 5 of the mold 6.

Referring to FIG. 6, the dimension of the orifice 11 measured axially is shown as T. In this figure, as in FIG. 7, the size of this dimension is enlarged for descriptive reasons. In an actual apparatus, the dimension T can range from about 0.01 to 0.075 inch depending on the characteristics of the polymer being formed and on the degree of orientation to be imparted. The orifice serves as the locus for high rate work input to the polymer that raises the temperature of the polymer to the orientation temperature range of the polymer, insuring good orientation characteristics. In general, the degree of orientation of the extrudate increases as the ratio increases between the average diameter of the extrudate as it emerges from orifice 11 and the average diameter of the slug.

The annular extrusion orifice 11 is area-convergent, as shown, in order to insure stable flow and a finite pressure drop between chamber 2 and the outer part of the orifice 11 during extrusion and especially at the time closing of the end of the bottle article is initiated; stated somewhat differently, a high pressure in the chamber 2 at the instant that the rod 4 is withdrawn, assures that

8

polymer will flow inwardly from the chamber 2 (with the continued urging of the ram 15), effecting a closure.

The mold cavity 5 has an annular groove 14 within its contour that is initially located adjacent to the discharge side of the annular extrusion orifice 11. The mold 6 with its mold cavity 5 can also be moved from the first location shown in FIG. 3 to a second location shown in FIG. 5.

Referring to FIGS. 1 and 2, the means for moving the various parts of the apparatus generally comprise hydraulic cylinders or hydraulic motors which are situated on a frame 25. The extrusion barrel 3 is flange mounted on a shelf 26 and is concentrically aligned with the hollow extrusion ram 15 which operates through an opening, not shown, in the shelf 26. Beneath the shelf 26 the ram 15 is aligned with and joined to the hollow tubular piston rod 28 of a non-differential, or double extended type, hydraulic motor 29 which is secured to the frame 25. Inside the bore of the extrusion ram 15 is the center supporting rod 4 which extends from the top of the extrusion rod 4 entirely through the extrusion ram 15 and the tubular piston rod 28. Beneath the lower end of the piston rod 28 the center supporting rod 4 is joined to the piston rod of another hydraulic motor 30 which is likewise secured to the frame 25.

At the upper end of the frame 25 is a dovetail slide 31 arranged to be moved parallel to the axis of the barrel 3 by means of a hydraulic motor 32 the body of which is secured to the shelf 26. Joined to the slide 31 is beam 33 which supports the mandrel 7 in axial alignment and spaced relationship with the extrusion barrel 3. Extend ing outward from the frame 25 is a fixed bifurcated stripping fork 34, the tines of which straddle the mandrel 7 directly beneath the beam 33. The mandrel 7 can be raised vertically by means of the motor 32 to effect the stripping of a formed container, not shown, from the mandrel. Additionally, this action exposes the extrusion chamber 2 in the barrel 3 to permit the insertion of a new slug of plastic.

Surrounding the mandrel 7 and the extrusion barrel 3 is a pieced mold 6 having apertures in its upper and lower extremities axially aligned with and slidable on both of these members. Mold 6 is movable vertically by use of hydraulic motor 35 the body of which is secured to the shelf 26. In FIGS. 1 and 2, the mold 6 is shown in its lowermost location, positioned for the initiation of a cycle. The mold 6 comprises two generally symmetrical halves with a planar parting face. The halves are hinged as shown in FIG. 1 but could be mounted on slides or links to permit the opening and closing on the extrusion barrel 3 and the mandrel 7. It should be understood that a movable clamping means, not shown, must be provided to secure the halves of the mold to each other to resist internal pressures of considerable magnitude. Such clamping means are well known in the art and generally comprise latches, pneumatic or hydraulic motors, screw clamps or the like. It will be understood that the mold walls may be made porous.

The mold parts may require heat or refrigeration depending on the material of the slug and may be provided with individual jackets or passages, not shown, for electrical or fluid heating or cooling. The extrusion barrel 3 may also require heat or refrigeration and jacket 36 is shown surrounding the part of the extrusion barrel that is accessible beneath the mold 6. If desired, the mandrel 7 can be equipped similarly.

Referring to FIG. 2, the hydraulic motors are controlled sequentially by means of several solenoid operated valves and an electrical control circuit. A gear pump 38 supplies fluid under pressure from a sump 39 to a plurality of conduits 40. The principal hydraulic motor is motor 29 which drives the ram 15 at a velocity determined by the profile of the cam 23. Cam 23 is carried on arm 27 by the rod 28 and is used to position a potentiometer 51 to produce a position-indicative output voltage. This output

3,733,309

9

voltage is fed to servo control 44 which, in turn, controls the operation of the valve 43 by varying the rate of flow of fluid to the motor 29 via valve 41 in proportion to the output voltage, the voltage and the fluid flow being higher when the high part 23b of the cam 23 is reached. The valve 41 is a self-centering four-way solenoid valve having ports which are blocked as shown when the solenoids are not energized. When the solenoid 42a is energized, the valve 41 swings clockwise admitting fluid to port 29a of motor 29 while simultaneously opening port 29b to "exhaust," thus permitting fluid in the upper part of motor 29 to return to the sump 39 via a conduit, not shown. The admission of fluid to port 29a causes the ram 28 to be driven upward. When the opposite solenoid 42b is energized, the valve 41 will move counter-clockwise and drive the ram 28 down again.

The arm 27 on piston rod 28 also drives the movable part of a potentiometer 24 which produces an output voltage proportional to the position of the rod 28 and ram 15 and varies in magnitude. This varying signal is used for controlling several events to be described. The system is activated by means of a power control circuit 45 which supplies electrical energy from source 46 directly to solenoid 42a on valve 41 and simultaneously to voltage comparison circuit 47 which also receives the input voltage from the potentiometer 24. The circuit 47 is adapted to produce three different outputs sequentially depending on the magnitude of the voltage output of the potentiometer 24 which depends on the position of the rod 28 of the motor 29. Thus, as the rod and ram move, the following events occur in sequence:

(1) The mold 6 is set in motion upward at a constant velocity by means of motor 35 via valve 49 which is actuated by solenoid 50a.

(2) A short time later, pressurized fluid is admitted via solenoid valve 52 and needle valve 53 to mandrel 7, passage 8 and ports 9 and 10 at a controlled rate of flow.

(3) Near the end of the stroke of the ram 15, the mold stops at the end of the stroke of cylinder 35 and the center supporting rod is then triggered into action by a signal from circuit 47 to solenoid 54a. This causes valve 55 to admit pressurized fluid to the upper port 30a of motor 30 thus causing the center supporting rod 4 to be pulled downward. It should be understood that the stroke of the motor 30 is very short, such as 0.1 to 0.2 inch. Thus, this motor and the rod 4 quickly "bottom" in the downward direction and maintain this position.

As the rod 4 moves downward, a lug on the rod engages limit switch LS-1 which then causes valve 56 to be energized to the opened position. This action bypasses the needle valve 53 and admits fluid into mandrel 7 and ports 9 and 10 at a greater rate than before. As an optional mode of operation of motor 30, as the limit switch LS-1 is closed, the solenoid 54b may be energized via switch 57. This causes valve 55 to turn clockwise, exhausting fluid from the upper port 30a of the motor 30 and admitting pressurized fluid to the lower port 30b of the motor. Thus, in a very short period of time rod 4 is pulled down, limit switch LS-1 is actuated and the rod 4 is urged upward again.

At the start of a cycle or at any time after a cycle has been completed, the mold 6 may be opened and the mandrel 7 may be withdrawn upward by use of motor 32. Valve 58 may be turned manually from the "rest" position shown in which pressurized fluid is admitted to port 32b thereby driving motor 32 and mandrel 7 upward to effect a stripping operation. At this stage, mold 6 and the ram 15 are in their uppermost positions. They are retracted by de-energizing the power control 45. This energizes solenoids 50b and 42b momentarily, causing valves 49 and 41 each to turn to admit pressurized fluid to the upper ends of motors 35 and 29, respectively,

10

and causing them to drive "down." The solenoids are then de-energized which permits the valves 49 and 41 to return to their "centered" position with all ports closed. The extrusion chamber 2 in extrusion barrel 3 is again ready to receive a fresh slug after which the mandrel 7 may be lowered by use of motor 32 and valve 58.

Precision control of the movable parts of the apparatus of this invention is achieved electronically by the control means shown in FIG. 2. For example, immediately after the formation of the bead in annular groove 14, the mold 6 is set in motion at a constant velocity when the voltage comparison circuit 47 senses a preselected level of voltage output at the potentiometer 24 when the ram 15 has gone through a preselected stroke of about 0.5 to 1.2 inches. In the first increment of mold motion, the annular bead moves along and over the end of the mandrel 7 causing the extrudate to cover the ports 9 and 10. With continued mold motion, the voltage comparison circuit 47 senses a different preselected level of voltage from potentiometer 24, triggering valve 52 and admitting fluid at a preselected rate through valve 53. This expands part of the neck portion of the extrudate outwardly against the mold surface in annular space 21 as shown in FIG. 3.

As movement of extrusion ram 15 and mold 6 continues, the neck portion is completed and the diverging part of the mold cavity 5 begins to pass beyond the region where the ports 9 and 10 are located, thus permitting newly extruded plastic to be expanded to a greater degree than in the neck portion as generally shown in FIG. 4. At this stage, when the polymer has reached, or is beginning to reach, the largest part of the mold 6, the ram 15 and rod 28 have advanced to the point where the high part 23b of the cam 23 drives the potentiometer 51 to a different position causing valve 43 to open to a greater degree. This admits fluid at a greater rate to motor 29 which increases the rate of extrusion of polymer through orifice 11 and contributes more polymeric material to the outermost wall of the article being formed.

As the ram 15 nears the end of its stroke, the potentiometer 51 returns once again to a lower level position on cam 23 so that the rate of extrusion of polymer through orifice 11 is decreased; this occurs substantially as the wall of the article is being completed and the bottom is to be formed and results in a thinner bottom.

Similar precision control is achieved just before the mold comes to a stop. The center supporting rod is triggered into action when the voltage comparison circuit 47 senses a preselected level of voltage from potentiometer 24 at which point solenoid 54a is energized and valve 55 operates admitting fluid to port 30a of motor 30. This starts to pull rod 4 away from abutment with the end of mandrel 7 and causes actuation of limit switch LS-1. This action triggers solenoid valve 56 thus by-passing valve 53 and admitting pressurized fluid to cavity 16 at a greater rate than before. This effects the completion of the expansion of the extruded shape in a shorter time than if valve 53 continued to control the rate of ingress of fluid.

Additionally, if desired, switch 57 may have been closed so that actuation of switch LS-1 would have the added effect of reversing valve 55 so that a very short time after its withdrawal, the rod 4 would have been thrust toward mandrel 7. The space formerly occupied by the tip of the rod 4 is then occupied by polymer which, by this action, is subjected to an impact squeezing effect in coordination with the urging of the ram 15. This mode of operation is preferred since the simultaneous actions result in a sound, high density closure.

An alternate mode of operation is particularly suited for polymers which, after drawing or orienting, exhibit poor conformability to a mold especially in the final stages of forming a container, in the present situation, the blowing and forming of the bottom of the container. This mode of operation comprises essentially the mode described above with an added step of reforming the bottom of the container inwardly to form a concave recess. This

3,733,309

11

is accomplished by the apparatus of FIG. 2 in which the voltage comparison circuit 47a is energized optionally by a switch 59; the circuit 47a also receives the voltage output of potentiometer 24 and is adapted to control "down" movement of the mold 6 via solenoid 50b of valve 49 in a manner to be described.

As described previously, travel of mold 6 in the upward direction during the forming of a container is controlled by voltage comparison circuit 47 which, when it senses that the output voltage of potentiometer 24 has reached a pre-set value, de-energizes solenoid 50a, restoring valve 49 to its "centered" position, thus stopping hydraulic motor 35. Ordinarily, the mold would continue to occupy the position at which it stopped; however, in the presently described mode of operation, at the instant that the solenoid 50a is de-energized, the voltage comparison circuit 47a (switch 59 closed), receiving the same voltage from potentiometer 24, acts to energize solenoid 50b; the latter actuates valve 49 to admit fluid to port 35b of hydraulic motor 35 thereby driving the mold 6 "down" again immediately. The voltage comparison circuit 47a after a short time delay de-energizes solenoid 50b, the mold 6 having moved through a short stroke (e.g., about 0.5 inch). When the solenoid 50b is de-energized the valve 49 becomes "centered" once again and the mold 6 comes to a stop.

The final steps in the forming of the plastic container, according to this mode of operation, are shown in FIGS. 8 and 9. In FIG. 8, the container is substantially complete with the mold 6 at the upward limit of its travel except that the plastic wall 60 has not conformed to the entire bottom portion of the mold, leaving a void 61 between plastic and mold. The next step, shown in FIG. 9, comprises moving the mold downward a short distance (e.g., 0.5 inch) to deform the plastic into a reverse bend and to form a truncated conical recess 62 in the bottom portion; during this step the generally conical wall 60' is stretched, imparting additional orientation to the conical wall 60'. It will be realized that the next and final step comprises close-off of the aperture occupied by the tip of the center supporting rod 4, a process step that has previously been described. The axial depth of the recess 62 or the downward stroke of the mold 6 is usually made great enough so that the plastic surface 63, FIG. 10, is either coplanar with the surface 64 or is slightly above it thus assuring stability when the container surface 64 rests on a support. The generally conical recess 62 serves the purpose of increasing the strength of the bottom of the container, improving its ability to resist internal pressures while minimizing the amount of plastic material needed for this purpose.

Referring to FIGS. 9 and 10, the bottom configuration of the bottle is described as a series of interconnected shapes beginning at the right cylinder section and terminating in the bottom center of the bottle comprised of (a) a curved section beginning at the right cylindrical section and curving in towards the longitudinal axis of the bottle forming a curve rotationally symmetrical about the longitudinal axis of the bottle and an annular seating area, that blends into (b) a truncated conical reentrant section leading into the interior, and directed towards the longitudinal axis of the bottle, that blends into (c) a toroidal knuckle forming an annular toroid directed away from the interior of the bottle, that terminates in (d) a central recessed disc-like section that is perpendicular to the longitudinal axis of the bottle.

In operation, the apparatus of this invention is used in the following manner: A thermoplastic polymeric slug 1 is placed within the extrusion chamber 2. Extrusion ram 15 is activated so as to force part of the non-molten thermoplastic polymeric material of the slug 1 through the annular extrusion orifice 11 and to the annular groove 14 in the end of the mold cavity 5. This first stage of extruding an annular bead from the thermoplastic slug 1 is shown in FIG. 3. It is seen that the first part of the slug

12

1 to leave the annular extrusion orifice 11 and enter the annular groove 14 forms a bridge or diaphragm around the entire upper part of the extrusion barrel 3 and the inside of the mold cavity 5 thereby effecting a seal. The extrusion of the slug into the groove 14 enables, in subsequent steps, the imposition of axial tension on the extrudate by moving the mold to stretch or draw the extrudate.

Immediately after the completion of the formation of the bead within the mold cavity 5 and in simultaneous sequence with the continued movement of the extrusion ram 15, the mold 6 is moved at a uniform rate of speed and a fluid, such as compressed air or liquid being packaged is forced through into the fluid passage 8, out of the fluid exit ports 9 and 10 and into cavity 16. This cavity is formed by the external surface of the mandrel 7, the extruded seal at the anular groove 14 and extruded shape 17 which was extruded by the annular extrusion orifice 11 and expanded by the compressed air from the fluid exit ports 9 and 10. This is shown in FIG. 4.

Thus, as mold 6 moves relative to the orifice 11, the bead formed in annular groove 14 anchors the newly formed bottle top to the mold 6 and effectively moves the fresh extrudate past the compressed air flowing from exit ports 9 and 10, thereby causing an almost immediate forcing of that extrudate against the wall of mold cavity 5 as it emerges from orifice 11.

The presently preferred method is centered on the production of a thermoplastic article having a nonuniform shell thickness due to the fact that the rate of extrusion and the speed of the mold are held constant while the mold itself has a varying shape. Shell thickness can be controlled by properly programming the apparatus to obtain either a uniform or a nonuniform thickness. Methods of programming shell thickness include varying the speed of the sliding mold or varying the extrusion rate of the slug.

The thermoplastic polymeric material of the slug 1 that is extruded through the annular extrusion orifice 11 becomes partially biaxially oriented from the extrusion operation. The remainder of the desired biaxial orientation of the extruded shape 17 is accomplished as the extrudate is drawn and expanded against the surface of the mold cavity 5 contained within the mold 6. There is a substantial decrease, e.g., up to 50% or more, in wall thickness of the extrudate after it has been drawn and expanded.

The slug 1 continues to be extruded through the annular extrusion orifice 11 by the extrusion ram 15 while the mold 6 moves toward the second location over the mandrel 7. The combined action of the extrudate 16 results in the desired shape of the bottle article 18 shown in FIG. 5, but having an unsealed bottom portion as best shown in FIG. 6. The bottom portion of the bottle article 18 is sealed by the withdrawal of the center supporting rod 4 while the mold 6 stops and the extrusion ram 15 continues to exert a force on the remaining polymeric material within the extrusion chamber 2. This is shown in FIG. 5 with the completely formed bottle article 19 which is in a highly biaxially oriented state.

FIGS. 6 and 7 show in greater detail the preferred bottom sealing operation in which the partial withdrawal of the center supporting rod 4 permits polymeric material of the slug, under the continued urging of the extrusion ram 15, to flow inward to effect a closure.

Alternatively, the bottom can be sealed according to the process disclosed in Carmichael, U.S. application Ser. No. 57,679, filed July 23, 1970, wherein a friction-welded bottom seal on a thermoplastic bottle is effected by contacting the bottom of the thermoplasic bottle in the area immediately adjacent the bottom opening with a rotating friction sealing head to raise the temperature of the thermoplastic material to about its melting point, working the hot thermoplastic material into and sealing the bottom

3,733,309

13

opening, and thereafter quenching the sealed opening. The process can be carried out while the bottle is still in the mold or in a separate operation after the bottle is removed from the mold.

FIG. 7 shows the location of the parts of the apparatus upon the completion of the method for forming a hollow article from a hollow slug. In FIG. 7 the center supporting rod 4 has been withdrawn while the extrusion ram 15 depressed the remaining portion of the thermoplastic slug into the volume vacated by the repositioned center supporting rod 4.

In an alternate embodiment, the apparatus is modified by providing a moving center rod through the slug instead of a stationary center rod. This permits the use of a closed end or blind hollow thermoplastic slug wherein the closed end becomes the bottom of the article formed, eliminating the need for a separate bottom closure step. In addition, the center rod and slug move at the same rate during extrusion of the slug eliminating relative motion between the slug and the center rod, thereby minimizing the need for lubrication between the center rod and the slug while reducing the wear in the center rod.

This alternate embodiment is shown in FIGS. 11 to 13 as it is used in the process of this invention. Referring to FIG. 11, a hollow cylindrical slug 1a having a closed end 65 is placed in extrusion chamber 2. Mold 6 is positioned in a first location wherein the mold cavity 5 surrounds extrusion chamber 2. Center rod 4a is positioned inside of slug 1a and extends into fluid passage chamber 8 within mandrel 7. If desired, the center rod can be heated by conventional means not shown, that in turn heats the thermoplastic slug. Extrusion ram 15a is modified to be a solid round bar positioned in extrusion chamber 2 abutting slug 1a containing the center rod 4a. The center rod is biased against the extrusion ram by conventional means not shown, exerting a nominal force against the extrusion ram sufficient to keep the extrusion ram from buckling the slug and to insure steady motion during extrusion. In a typical embodiment, the extrusion ram pressure is about 13,000 p.s.i.g. and the center rod biasing pressure is about 50 p.s.i.g.

In operation, extrusion ram 15a forces slug 1a out of extrusion chamber 2 through extrusion orifice 11 and around mandrel 7. FIG. 12 shows an article partially formed, wherein in the extrusion ram 15a has forced the slug into the sliding mold while simultaneously moving the center rod with the slug so that there is no relative motion between the slug and the center rod. Fluid is introduced through fluid passageway 8 around center rod 4a and out through port 10 into the interior portions of the extruded shell 17, forcing it against the mold cavity thereby shaping the article.

FIG. 13 shows an article completely formed inside of mold cavity 6. It can be noted that the bottom of the slug 65 is now the bottom center portion of the article. In addition, that portion of the center rod 4a that originated in the slug while the slug was in the extrusion chamber has been moved into fluid passageway 8.

After the thermoplastic article is formed, it can be heat treated by well known processes to increase the crystallinity level, thereby decreasing the ability of the gases to permeate the shell and improving dimensional stability which is important if the article is used to bottle hot beverages or is to be subjected to high temperatures and pressures in a pasteurization process.

Heat treatment is carried out at temperatures of about 140°–220° C. and exposure time is relatively short. However, it is generally desirable to conduct the heat treatment for a period of time sufficient to produce a degree of crystallinity in the finished product which is preferably at least about 30% up to 50% or more, the maximum attainable crystallization for polyethylene terephthalate being about 60%. In general, especially good results have been observed when this heat treatment step is carried out for a period of about 0.1 to 600 seconds. The upper

14

limit of this treatment is not particularly critical, other than from an economical viewpoint, and a duration of treatment of up to 100 minutes is possible.

The thermoplastic slug useful in the present invention is hollow, but the term "hollow," unless otherwise indicated, is meant to include a tubelike slug having both ends open, or a tubelike slug having one end open and one end closed, i.e. a blind slug, wherein the slug is so positioned in the extrusion barrel that the closed end will form the bottom of the bottle. The tubelike slug having both ends open can be used with the apparatus incorporating a stationary center rod or a moving center rod but a blind slug can only be used with the apparatus incorporating a movable center rod.

The slug is preferably fabricated by conventional extrusion or injection molding methods from thermoplastic materials which are susceptible to increased strength or reinforcement when biaxially oriented. The slug itself can be biaxially oriented or unoriented prior to use. If an oriented slug is used, further orientation occurring in the extrusions drawing and expansion of the extruded slug is additive in effect. In addition, the slug should be practically amorphous with no more than about 5% crystallinity and clear in appearance. This will result in a clear formed bottle. If it is intended that the bottle be colored, however, the coloring agent such as a dye can be added to the slug forming polymer and, of course, result in a colored slug.

The dimensions of the slug to be used are determined by many factors including the desired thickness and the desired degree of orientation. Typically the slug is hollow and the radial dimensions are slightly smaller than the dimensions of the neck of the bottle to be formed as can be seen in the drawings. The axial length of the slug is slightly shorter than the dimension between the top and the center of the bottom as measured along the outside of the bottle to be formed. To improve the dimensional stability of the bottle, particularly the radial dimensions of the neck of the bottle, the slug first formed with substantially oversized radial dimensions, quenched to a temperature below the crystalline melt point of the polymer then forced through a reducing die slightly smaller than the radial dimensions of the neck of the bottle as shown in the drawings. For still further improved dimensional stability, the slug can be compressed in a chamber maintaining the same outside diameter with a tapering mandrel in the center of the compression chamber resulting in a very short slug having an outside diameter slightly smaller than the outside diameter of the neck of the bottle and an inside diameter of practically zero resulting in a very narrow hollow space about the size of a pin hole running through the center of the slug. The compressed slugs are used in the apparatus described above without the presence of the center rod or with the center rod fully retracted.

The process and apparatus of this invention can be used to prepare articles of various shapes and sizes from various thermoplastic materials. The preferred thermoplastic material is polyethylene terephthalate, and copolymer blends thereof as described above.

One reason polyethylene terephthalate is preferred is because when oriented, it exhibits excellent strength, creep resistance, and a low permeation factor, particularly with respect to carbon dioxide, oxygen and water vapor, making it excellently suited for use as a container for liquids bottled under pressure, such as sodas, beer, or aerosols. When forming with polyethylene terephthalate, it is advantageous to start with essentially amorphous material, i.e. crystallinity no greater than 5% in order to produce a clear bottle. Useful polyethylene terephthalate polymers have an inherent viscosity (1% concentration of polymer in a 37.5/62.5 weight percent solution of tetrachloroethane/phenol, respectively, at 30° C.) of at least 0.55. Preferably, the inherent viscosity is at

3,733,309

| 15 | 16 |

least 0.7, because this produces a bottle having significantly improved toughness properties, e.g. increased impact resistance.

Impact resistance is measured by dropping a slug from various heights onto a concrete floor. In a drop test carried out on six-inch-long, amorphous polyethylene terephthalate having an inherent viscosity of about 1.1 wherein three slugs were used in testing having an average wall thickness in mils of about 138, 90, and 93, with a weight in grams of 27.8, 21.2, and 21.6, respectively, each slug sustained two drops from a height of one foot, two feet, five feet and eight feet without any apparent damage to the slug and, in addition, each slug sustained the impact of a five-pound weight dropped two times onto the slug from a height of one foot.

Other useful thermoplastic materials include copolymers of acrylonitrile/styrene/acrylate; acrylonitrile/methacrylate; methacrylonitrile copolymers; polycarbonates; polybis(para-aminocyclohexyl) dodecaneamide and other polyamides; polyformaldehyde; high density polyethylene and polyproplylene; other polyesters and polyvinyl chloride.

Laminar-walled bottles or the like can be produced by the process of this invention by employing a laminar-walled hollow cylindrical slug. Laminar-walled slugs are obtained by coaxially laminating two or more slugs of the same or different thermoplastic composition. Examples of practical combinations include polyethylene terephthalate on the inside coaxially laminated to polyvinylidene chloride copolymer or hydrolyzed ethylene vinyl acetate copolymer on the outside. Slugs of multi-polymer composition can be coextruded in two or more layers, i.e., preferably in three layers, with the additive polymer sandwiched between the base or bottle-making layers of polymer. Through the use of such a slug, it is possible to produce bottles of base resins with a selected laminate to be used as (1) a gas barrier, (2) coloring layer, or (3) degrading catalyst.

The extruded slug must be at a temperature within its range of biaxial orientation, i.e. the temperature range for the polymer being used wherein orientation can occur without line drawing. The heat generated during extrusion is generally sufficient for this purpose so that the slug can be extruded at room temperature. However, the orientation temperature range varies from polymer to polymer, depending on such factors as crystallinity, and the glass transition temperature of the polymer. If the orientation range of the polymer is so high that the heat of extrusion is not sufficient to raise the polymer temperature to its orientation range, then the slug can be preheated before extrusion.

The thermoplastic article formed is biaxially oriented and will have physical properties consistent with the type slug used.

The following examples illustrate the present invention. All parts, percentages and proportions are by weight unless otherwise indicated.

EXAMPLE 1

Polyethylene terephthalate polymer of inherent viscosity about 0.96 is made into a hollow cylindrically amorphous shaped slug 4.5 inches long, 0.680 inch outside diameter (O.D.) and 0.375 inch inside diameter (I.D.) weighing about 22.6 grams. The slug is preheated to about 92° C. and extruded through gap T of about 0.033 inch at a barrel temperature of about 85° C. in the apparatus described above. The velocity of the ram 15 is about 3.6 inches per second and the velocity of the mold 6 is about 5.1 inches per second. Air at about 255 p.s.i.g. pressure is introduced through ports 9 and 10. The internal mold diameter is about 2.5 inches.

A bottle is formed having a wall thickness of about 11.4 mils; the axial tensile strength is about 16,500 p.s.i. and the hoop tensile strength is about 26,700 p.s.i.

EXAMPLE 2

Example 1 is repeated except as follows:

Inherent viscosity	1.0.
Slug length	6.5 inches.
Slug O.D.	0.680 inch.
Slug I.D.	0.477 inch.
Slug weight	23.5 grams.
Preheat temp.	100° C.
Barrel temp.	90–100° C.
Gap	0.035 inch.
Ram velocity	5 in./sec.
Mold velocity	5.8 in./sec.
Air pressure	350 p.s.i.g.
Tensile (axial)	8,000 p.s.i.
Tensile (hoop)	30,300 p.s.i.
Wall thickness	16.8 mils.

A thermoplastic bottle is produced according to the

EXAMPLE 3

procedure of Example 1, namely extruding and blow-molding a hollow cylindrically shaped slug 4.5 inches long with an outside diameter of 0.680 inch and an inside diameter of 0.375 inch weighing about 22.6 grams. The slug is made from polyethylene terephthalate which has a inherent viscosity of 0.91. The slug has a density on the outside surface of 1.332 and on the inside surface of 1.334, and a crystallinity of about 5%.

The bottle exhibits the following properties:

I

Density and crystallinity of polymer from various points on the bottle	Density	Crystallinity, percent
Neck	1.332	0
Top of major cylindrical section	1.345	6
Middle of major cylindrical section	1.356	17
Bottom of major cylindrical section	1.361	22
Bottom of bottle	1.332	0

II

Tensile properties, right cylinder section	Axial	Hoop
Tensile strength (k.p.s.i.)	7.8	23.8
Elongation (percent)	59	17
Tensile modulus (k.p.s.i.)	246	683
Yield stress (k.p.s.i.)	7.6	10

III.—BIAXIAL ORIENTATION RIGHT CYLINDER SECTION
[X-Ray orientation angles according to the articles incorporated by reference above]

2θ peak	Direction of rotation, X(chi), φ(phi)	Orientation angle	Peak max.
17.0	Plane perpendicular to beam	83 (axial)	0° X
	Plane parallel to beam, scan 90	52 (hoop)	0° X
	0 scan	46 (hoop)	0° X
27.0	Plane perpendicular to beam scan 0		
	Plane parallel to beam, scan 90	32 (hopp)	8° X
	0 scan	40 (hoop)	87° X

Considering the X-ray orientation angles and the tensile properties above, the bottle exhibits and effective stretch ratio of about 3.5 times in the hoop direction and about 1.25 times in the axial direction.

(IV) PERMEABILITY RIGHT CYLINDER SECTION

Shell thickness ---------------- 18 mils.
Water loss—Bottle filled with
water stored at 17.5% relative
humidity, 25° C. for 13 days ___ 0.6 mg./hr.
Carbon dioxide loss—Bottle pressurized with carbon dioxide to
a pressure of 40 p.s.i.g. at 25° 1.5 cc./day.
C. Bottle had no permanent de-   (standard temperature and pressure).
formation ------------------

(V) CREEP

Circumferential strips from the right cylinder section of a sacrificed bottle at 50° C. resist a hoop tensile stress of 5,000 p.s.i. with creep at 100 hours at a value less than

3,733,309

17

2% and long term creep of 90 days, less than 5%. This corresponds to a deformation constant of about 1.5.

As many widely different embodiments of this invention may be made without departing from the spirit and scopet hereof, it is to be understood that this invention is not limited to the specific embodimens thereof except as defined in the appended claims, and all changes which come within the meaning and range of equavilence are intended to be embraced herein.

We claim:

1. A biaxially oriented bottle having a generally cylindrical section wherein the bottle has a density of about 1,331 to 1,402 and is prepared from polyethylene terephthalate having an inherent viscosity of at least about 0.55 with a ratio of the weight in grams of the polyethylene terephthalate to the inside volume of the bottle in cubic centimeters of about 0.2 to 0.005:1, and the generally cylindrical section of the bottle has an axial tensile strength of about 5,000 to 30,000 p.s.i., a hoop tensile strength of about 20,000 to 80,000 p.s.i., an axial yield stress of at least 4,000 p.s.i., a hoop yield stress of at least 7,000 p.s.i., and a deformation constant equal to the slope of the log (reciprocal of the strain rate) versus strain having a value of at least about 0.65 at 50° C.

2. The bottle of claim 1 in which the inherent viscosity of the polyethylene terephthalate used to prepare the bottle is at least about 0.7.

3. The bottle of claim 1 having a crystallinity in the generally cylindrical section of the bottle of at least about 15%.

4. The bottle of claim 1 prepared from polyethylene terephthalate mixed with a pigment.

5. The bottle of claim 1 having a shell thickness of about 10–30 mils in the generally cylindrical section of the bottle.

6. The bottle of claim 1 wherein the crystallinity of the polyethylene terephthalate is at least about 15% and the carbon dioxide permeation is no greater than 15% in 30 days when the bottle is at room temperature and contains about 75 p.s.i.g. autogenous pressure.

7. The bottle of claim 1 wherein the crystallinity of the polyethylene terephthalate is at least about 15% and the oxygen permeation infusion from the atmosphere into the bottle filled with a liquid and containing 75 p.s.i.g. autog-

18

enous pressure at room temperature is no greater than 5 parts per million in the liquid in 30 days.

8. The bottle of claim 1 wherein the crystallinity of the polyethylene terephthalate is at least about 15% and the water moisture permeation from the bottle containing water at room temperature and under an autogenous pressure of about 75 p.s.i.g. is no greater than 5% in 90 days.

9. The bottle of claim 2 wherein the generally cylindrical section of said bottle has a crystallinity of at least about 15% and a shell thickness of about 10 to 30 mils, and bottle permeation characteristics wherein the carbon dioxide permeation is no greater than 15% in 30 days when the bottle is at room temperature and contains about 75 p.s.i.g. autogenous pressure, the oxygen permeation infusion from the atmosphere into the bottle filled with a liquid and containing 75 p.s.i.g. autogenous pressure at room temperature is no greater than 5 parts per million in the liquid in 30 days, and the water permeation from the bottle containing water at room temperature and under 5% in 90 days.

10. The bottle of claim 9 prepared from polyethylene terephthalate mixed with a pigment.

11. The bottle of claim 1 in which the inherent viscosity of the polyethylene terephthalate used to prepare the bottle is at least about .90.

12. The bottle of claim 1 in which the inherent viscosity of the polyethylene terephthalate in the bottle is at least about 0.85.

13. The bottle of claim 1 in which the inherent viscosity of the polyethylene terephthalate in the bottle is at least about 0.95.

14. The bottle of claim 1 in which the inherent viscosity of the polyethylene terephthalate in the bottle is about 1.1.

References Cited

UNITED STATES PATENTS

3,294,885   12/1966   Cines et al.
3,314,105   4/1967   Amsden.
3,470,282   9/1969   Scalora.

MELVIN GOLDSTEIN, Primary Examiner

U.S. Cl. X.R.

18—5 BE; 215—1 C; 264—98, Dig. 33, 50, 64

ADDITIONAL READING

GC Heldrich. The history of blow molding. SPE J 435, 1961.

Modern Plast pg. 5, 1949.

Blow molding, Modern Plast, pg. 83, 1969.

J Bailey, Blow molding, Modern Plast 127, 1945.

D Schmidt. Blow molding fundamentals. Modern Plast pp. 105 & 189, 1961.

WD Bracken. Blow molding. Technical Papers, Vol. VI, 16th Antec, SPE 1960.

The Plastic Blow Molding Handbook, SPE pp. 12 & 14, 1991.

M Bakker, Plastic bottles and changes in the big picture, Seventh
  Annual Blow Molding Conference, SPE Blow Molding Division,
  Oct. 1991, p. 8.
RE Dunham, Technical history of blow molding, SPE, Antec 1992.
Plast World 581, 1993.

2

SELECTING THE PROPER EQUIPMENT FOR INDUSTRIAL/TECHNICAL BLOW MOLDING

Robert A. Slawska

Proven Technology, Inc.
Belle Mead, New Jersey

Blow molding allows producing small to large, hollow, double-wall parts that require exceptional rigidity, heat deflection, flame retardance, toughness, and impact strength. Many of these parts are not practical to produce by injection or structural foam molding.

The following are markets for industrial/technical blow molding:

Automotive
Recreational/ornamental
Toys
Cooperage
Hospital supplies
Housewares

Household appliances
Furniture

Typical products produced in these industries include:

Fuel tanks
Bumpers
Spoilers
Seating
Porta-toilets
Storage boxes
Chests, coolers, food trays
Playhouses
Gym sets
Drums
Doors
Panels
Filing cabinets
Computer workstations
Tool cases
Game cases
Furniture parts

2.1   HOW PARTS ARE BLOW MOLDED

First, the plastic is melted in the extruder. Then, the plastic is formed in the basic head as a preform annulus. The hot plastic parison is rammed out and trapped by two mold halves on a clamp station with high-pressure lockup. Air is injected at around 100 psi into the inside of the parison. This forces the parison to inflate and take the form of the mold cavity. The mold is chilled to around 50°F. Then, the cold surface is transferred onto the plastic surface to help to cool the part being made. When the plastic is set up, the enclosed air is vented (exhausted) from inside the part. The mold opens and allows the part to be removed from the clamp station. Then, the cycle repeats (Figs. 2.1 and 2.2).

Figure 2.1 Hartig blow molder with 100-pound shot, two 6-inch (150m) 24:1 grooved feed extruders with a 300 ton clamping system. (From Davis-Standard, Edison, New Jersey.)

2.2 THE POSITIVE BENEFITS OF THE PROCESS

With blow molding, large parts can be produced independently of mold flow or large clamping capacity associated with injection or structural foam molding. Blow molding has the lowest tooling cost except for vacuum forming and 35% of

Figure 2.2 Wilmington Machinery Model Blow Molder with three re-ciprocating extruders for multilayer products. This unit is producing a 32-gallon multilayer trash container.

injection molding. It also has the shortest tooling lead times except for vacuum forming. Integration of inserts permits molding many different parts from one mold. Double-wall configuration provides increased modulus with less material, less weight, and the best strength to weight ratio of all comparable processes.

2.3 NEGATIVE FACTORS OF ACCUMULATOR HEAD BLOW MOLDING

Blow molded parts are made by forming a round parison. This is usually satisfactory for producing round parts. However, many new products are being developed, such as furniture,

that is rectangular and has a thin wall. Forming these types
of parts require large parisons and prepinch/preblow systems.
They are difficult to duplicate consistently. Usually, the walls
of the parison that come in contact first with the molds produce thick sections. The extreme outside wall of the parison is
stretched as the molds are touched. This results in thinning of
the parison wall and hence the finished parts. The thin portions produce sections that will be out of spec. The thick sections take longer to cool and increase cycle times. This can
also distort the part if removed too soon from the mold. In
general, the following are major problems in typical accumulator head blow molding:

> Round parisons are used to make rectangular parts.
>
> Cross sections are produced with thick and thin irregular
> wall sizes.
>
> Constant weights are difficult to maintain due to the pre-
> pinch/preblow process.
>
> Die lines in finished parts.
>
> Pleating of parison causes heavy sections that are over-
> lapped.

Industrial blow molding equipment is becoming much more
scientific. The equipment and the process is more reliable and
predictable due to design improvements and microprocessor
advances in operating the basic equipment.

The seven major parts of the basic machine listed following must be coordinated to meet product specifications. There
are many alternatives in selecting the proper equipment.

Component	Requirements
Extruder	Output and material
Accumulator head	Capacity and material
Press	Tonnage and clear platen area
Hydraulic unit	Tank size, pump, and hp
Pneumatic system	Size and number
Microprocessor	Type and ease of setup
Takeout system	Controls and type

2.4 HOW TO SELECT THE PROPER EQUIPMENT

In the early stages of selecting the proper equipment, it is necessary to identify the part or parts to be produced on this equipment. It must be determined exactly what is required:

Products (full range)
Sizes
Quantity
Material to be used
Finished weight of part
Wall thickness-minimum/maximum
Cycle time
Future products

For each part to be made on this machine, we should estimate certain parameters to allow appropriate equipment size. Also consider what might be produced in the future after completing this project. This is necessary in the case of newly designed parts with no prior history of how they will be produced. We will want to estimate the following information:

Head tooling size
Minimum wall thickness of part
Length of parison
Shot size and weight
Finish part weight
Simple parison programming profile
Flash requirements:
  Top and bottom
  Sides, preblow, and pinch

Now is the time to develop the throughput versus process parameters for a simple blow molded case as an example:

3600 (seconds) divided by estimated cycle (seconds) = Yield.

Assume: 70 second cycle. Therefore,

3600/70 = 51 parts/hour (Yield) per cavity.

Assume: Seven pound shot size (3178 grams).

The formula is Yield × number of heads (H) × weight (shot size) = Output: Y × H × W = O

$51 \times 2 \times 7 = 714$ lb/hr

2.5 THE EXTRUDER

The extruder is used to deliver a homogeneous and stable melt to the accumulator head. The extruder is truly the heart of the melt preparation system for blow molding. Typical extruder sizes for blow molding range from 2 1/2 in. to 6 in. in diameter. The L/D ratio is normally 24:1. Listed below are the operating specifications for various extruder sizes when using HDPE whose melt index is 0.2.

Extruder size (in.)	Power (hp)	Output (lbs/hr)		Screw speed (max rpm)
		at 100%	at 80%	
2 1/2	50	300	240	125
3	75	400	320	125
3 1/2	100	600	480	125
4 1/2	200	1000	800	80
6	300	1500	1200	70

These data are based on a smooth bore, feed throat section. This design is an industry standard. Now, grooved feed throats are becoming very popular for HDPE. They were developed to be used with HMWHDPE. The grooved throat feature actually increases the outputs listed by approximately 25 to 35%. This increased output is accomplished with a lower screw speed and a lower melt temperature and results in a faster cycle time. The grooved feed throat provides a positive feed of the plastic pellets into the feedscrew.

Based on our previous example, 714 lb/hr are required to produce the parts. Using the chart, a 4 1/2 in. extruder has an output of 1000 lb/hr. This would be the logical choice and allow for future output expansion if a faster cycle time can be developed. At 80% performance, this 4 1/2 in. extruder still has an 800 lb/hr output. During normal operation, the frictional heat from the feed screw revolving within the barrel creates 80% of the heat developed in the process. It is much preferred to run the extruder at lower speeds to produce the desired output.

The two most popular methods of cooling the extruder barrel are by closed-loop water cooling or air-cooling blowers. The extruder is divided into various heated and cooled zones. A 4 1/2 in. extruder has five zones. These are controlled automatically by the microprocessor. Air-cooled barrels have blowers with very high cfm air flow capacity. Today, most of the extruders up to 4 1/2 in. are cooled by air blowers. Closed-loop cooling water systems are usually used on 6 in. extruders. Closed-loop systems can be supplied on any size extruder.

Serious thought should be given to the gear reducer. The gear reducer decreases the normal extruder motor speed range of 300 rpm to 1750 rpm to a workable screw speed of 0 to 125 rpm (3 1/2 in. extruder). A light duty gear reducer will not allow continuous operation with today's high torque materials. Gear reducers have excellent service factors at high screw speeds. At lower speeds, the service factor is rapidly reduced. The ideal service factor for a gear reducer is 1.50 at normal operating speed. Make sure that the thrust bearing and radial bearing are sized to match the horsepower and torque ratings of the extruder. These bearings will provide trouble-free and long, continuous operation. Hence, the service factor is actually the safety range in which a gear reducer can perform within a defined speed and horsepower.

The extruder for blow molding is extremely important because a wide range of materials and outputs may be used regularly. The most commonly used materials in industrial blow molding machines are high density polyethylene and polypropylene. A large percentage of polypropylene blow molded parts are used in automotive applications. There is a growing usage of ABS and other engineering materials. Regardless of what material is to be used, a highly critical consideration is the ability of the extruder to deliver a homogeneous and stable melt to the die head. If this is not accomplished, it is extremely difficult, if not impossible, to consistently produce acceptable blow molded parts.

Many existing blow molding machines have feed screws improperly designed for the specific material used. In fact, screw designs are seldom discussed when considering equipment for blow molding applications. This should be one of the first areas to be considered in attempting to set up and run an efficient blow molding operation. Once the melted material is in the accumulator head, it is too late to complete the necessary melting or provide a homogeneous melt. The result is an unstable parison in terms of length, size and temperature. This leads to poor part appearance and inferior products.

Proper screw design, on the other hand, prevents any of the solid mass (pellets) from discharging into the melt pool. Proper screw design also requires less shear on the material extruded to the head, resulting in definitely lower melt temperatures and proper mixing and dispersion. So this is the first step in good parison formation and a major move toward achieving lower overall cycle times.

Screw speed should be variable to allow for a wide range of resins, parts, and operating needs. Variable speed control also gives fine adjustments in output. The most commonly used extruder drive system is a SCR type dc package. This system converts ac voltage into dc voltage. This type of system has been commercially available for many years and most of the bugs have been worked out. Variable frequency ac drive systems are now being considered as a replacement for the dc drive. This system is very efficient, easy to maintain, and has energy savings over a dc drive. The ac motor does not require a blower or filter and is also very quiet.

2.6   ACCUMULATOR HEAD

An accumulator head is used to store a certain amount of melted plastic material developed by the extruder and transferred into the accumulator head through a melt transfer adapter. The head creates a preformed annulus under low extrusion pressure while filling. The accumulator head can produce various shot sizes based on the total capacity of the head. Typical head sizes are shown below. These heads are rated by full shot capacity in pounds (HDPE 0.2MI).

Size (lb)	Tooling size range (in.)
1	1/2–4
2	1–6
3	1–6
6	2–9
8	2–9
10	2–11
15	2–14
20	2–14
25	5–18
35	5–18
40	5–18
50	6–24
75	6–24
100	10–36

It is desirable for the accumulator head design to have a first-in, first-out material flow. The head design is the most important part of a blow molding machine. A poor head design always gives poor parisons and usually marginal or bad parts. When designing these types of heads, the following items must be considered:

Solid rod to annulus formation	Bleed out/lubrication methods
Material sticking to walls	Parison formation
Streamlining flow surface	Speed of parison drop
No decompression areas	Parison programming
No reknit traces in parison	Head tooling design
No stagnant areas	Parison swell
Excellent parison concentricity	Head filling

The extruder produces a solid rod discharge like a pipe into the transfer adapter. This solid rod must be formed into a preformed annulus (parison). This is very difficult to accomplish. Most of the previous considerations must be analyzed for proper head performance. While selecting the shot capacity of the head, care should be taken not to use an oversized head. It is not recommended to use a five pound shot on a 40 pound head. This could develop stagnant flow areas that create dead spots within the internal flow surfaces of the head which will allow the plastic material to burn and produce black specks on the finished parts, resulting in rejects. Clean-out of heads will take longer by using less inventory (shot capacity) of the head on a continued basis.

2.7   IMPORTANCE OF CLEANING

It is desirable to be able to change colors or materials very quickly in a blow molding machine. Unfortunately, the changeover time can range from a few hours to as many as 30 hours. The cost of changeover ranges from $700 to more than $6000 per change. The net result is lost profits due to nonproductive time.

Most changeovers are accomplished by continually purging of natural material and finally adding the new color or material.

Another consideration is disassembling, cleaning, and reassembling the head. In most cases, this takes from eight hours to three full days depending on the head and whether it is completely disassembled in the flow areas (Fig. 2.3).

Of course, machinery builders will not admit that changing colors or material in their heads is a problem. A perfect solution is to actually do testing on the equipment to determine the actual changeover time. This should be done before purchasing any new equipment.

When determining the tooling size, the actual outer diameter of the main body will set the possible range. If the body has an o.d. of 20 in. for a 25 pound head then the maximum tooling would be 18 in. in diameter. If a larger head tooling is needed over standard, the programming cylinder may not have the capacity to control the movement. The program cylinder must be at least 30% more powerful than the shooting cylinders. With diverging tooling, more power is required to efficiently control programming.

Parison swell will be largest with small converging tooling, say 4 in. in diameter. The swell at this size is around 40%. With 14 in. convergent tooling, the swell ratio is only around 20%. This is based upon HDPE. When using engineering material, the actual swell ratio may be only from 10 to 18% within the tooling range.

Die swell is a very complex function. It occurs as the parison is being pushed out of the head tooling. There are two types of swell. One is diameter swell. This is easy to see because the parison actually balloons in size and is bigger than the die. The other type is weight swell. This occurs as the parison drops and is difficult to see. Actually the wall thickness of the parison increases as the length shrinks. Diameter swell is easier to calculate than weight swell. This swell ratio differs from head to head and manufacturer to manufacturer. Listed below are some typical parison diameter swell increases:

Figure 2.3 Graham Engineering 20-pound accumulator head. Close up view of diverter and plunger retracted for inspection and cleaning. Notice access around diverter and inside of head body.

Material	Percent
HDPE	25–65
LDPE	30–65
PVC	30–35
PS	10–20
PC	5–18

The land length and die gap opening play an important part in swell. Following is a suggested land length for a given annulus die gap:

Annulus gap (in.)	Land length (in.)
Above 0.100	1–2
0.30–0.100	3/4–1
Below 0.030	1/4–3/8

Many factors affect die swell. In fact, this is such a complex topic that a complete book or chapter could be written on the subject. This author will not cover this. Different heads do not provide the same parison swell. Every machinery manufacturers' head performs differently. It is best to determine your actual part lay flat requirements to establish whether the head will give you the proper size parison to make the part.

Changing head tooling is a tedious job that takes several hours. Consideration should be given to the method of changing tooling. If tooling is not changed often, then there is little to be concerned with. However, most molders change tooling very often. Quick change tooling should be used, and several types are available. Split two-piece tooling is the most common and the most economical. This allows for removing only the bottom half of the tooling and is quicker and less expensive. There are also quick change die ring systems available, which allow faster changing of the die bushing.

It is recommended that the machine have the ability to control push-out speeds and fill rates of the accumulator head. There are several excellent parison programming systems available. This should be covered in a different chapter. The parison programmer should have rapid response and at least 100 program steps.

Most of the machines that produce up to 25 pound heads are set up for dual-head operation (over 70%). Each head should be independent from the other. The system must consistently control the fill rates from cycle to cycle. If not, the reject rates will be very high. The system must also control uneven shot sizes during dual-head operation. This must be an automatically controlled, and the fill rate must vary less than 1% from shot to shot and cycle to cycle. Parison final lengths must be consistent from cycle to cycle and head to head.

By determining the actual cycle time and productivity yield, you can establish whether a single head or dual head operation is required. A dual-head operation will almost double your output based upon the part removal system. The actual price increase from a single-head machine to a dual-head unit is approximately 35 to 40%. However, productivity doubles when using dual heads (Fig. 2.4).

2.8 PRESS

The press forms the blow molded part. The molds are mounted to the platens. Determine the actual size of the molds to be used. This should include outriggers, such as split mold mechanisms and cylinders, needle blow cylinders, water lines and clamping devices. Enough room should be left on the platens to allow for these items. If there are two molds (dual heads), then there must be enough space between both molds to allow for cooling lines and cylinders. It is important to determine the actual head centerline spacing. It is best to have the actual dimension when the heads are fully heated, because they expand when up to temperature. This allows the molds to be mounted on the proper centers.

It is very important that the daylight minimum and maximum openings be readily and simply adjusted to compensate for various mold sizes and shut height. Backup plates are used with dual molds that make mounting to the platens very easy. Make sure the daylight opening is large enough to remove the part from the mold. This also helps to maintain

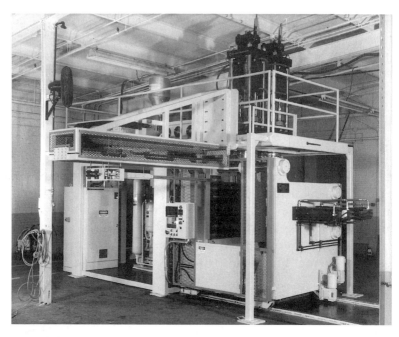

Figure 2.4 Graham Engineering Model GEC-10DP6040. Dual-head operation provides up to 10 pounds shot at each head.

the necessary head center distance. The platen closing must be synchronous and parallel to produce acceptable parts. If not, excessive wear occurs on the guide rod bushings, resulting in reject parts. Blow pin damage is also common due to uneven closings. The closing of the platens must be smooth, without hesitation or bounce-back, especially during the final closing. If not, bad pinch-off results. Most press designs use guide rods to mount the platens. There are platen designs that do not have guide rods. The presses are mounted on slide ways for movement and alignment.

There are two types of clamp forces, One for pinch force and the other for blow force. Pinch force occurs first and then blow force. These forces are not combined and occur one after the other. To determine pinch force, the force per linear inch area must be determined. Add up all the pinch area over the entire mold. Determine the type of plastic material to be used. HDPE has a force requirement of 400 to 500 pounds per linear inch. PP copolymer requires a force of 600 to 700 pounds. These are based on a pinch land of 0.015 in. The simple formula to determine pinch force is length of pinch area multiplied by pounds of force per linear inch divided by 2000. To determine blow force, use the projected area of the blown section multiplied by the air blow pressure divided by 2000. By using both of these simple formulas, the correct clamp tonnage can be determined for the parts to be produced.

The design of a press should include the following considerations:

Platens must be flat.

The platens must be parallel and square face-to-face.

No protrusions—must have clean faces.

Size platens to fit molds (width and height without overhanging molds).

Proper clamp tonnage. Compare parts to be produced to determine tonnage.

Platen speed control with microprocessor ramp.

Maximum platen deflection of 0.006 in. on 36 in. × 36 in. length and height.

Accessible mold mounting

Accessible part removal.

Motorized vertical platen adjustment.

Motorized rollout for mold changeover.

Platens must close on positive centerline.

No shock on platen hydraulic movement.

Instantaneous high-pressure lockup.

Convenient platen hole layout or "T" slots.

Platen speeds up to 1200 in./min with programmed position points.

Automatic operator safety gate with hydraulic and electric safety.

Adjustable high-pressure tonnage.

2.9 HYDRAULICS

Hydraulic systems are rapidly advancing in design and capabilities. Hydraulics is the most difficult part of the system to understand, especially, when it is combined with the knowledge of microprocessors to make it function. These systems use proportional valves to control must of the hydraulic functions. Now, the microprocessor uses closed-loop control for the valves. With the ramp feature of the microprocessor, we can consistently and repeatably control the positions required by the blow molding unit. New systems, such as regenerative on the press cylinders, are now being used. This enables using smaller horsepower motors and less oil with smaller reservoir sizes. All of these factors greatly improve the operation of the blow mold unit. There must be better oil filtration and maintenance of the hydraulic system to maintain these levels of improvement. Smaller motors are energy savers.

Variable displacement pumps are replacing fixed vane pumps. These pumps save energy and are more efficient. However, because variable pumps and proportional valves are, the hydraulic oil must be finely filtered to 5 microns. You cannot see the oil impurities that are less than 40 microns. A well-designed hydraulic system should include the following:

> Properly sized heat exchanger
> Clean out covers on the tank
> Drain plugs
> Suction strainers mounted inside the tank
> Manual shut-off valves for each section
> Automatic dump valve for supplied nitrogen bottle
> Noise levels lower than 85 decibels
> Accessibility of tank and components

2.10 MICROPROCESSOR

Industrial blow molding is still considered an art, not a science. The advent of microprocessors has gone a long way in developing the state of the art in blow molding. Since 1979, these systems have developed into useful, successful controls. These are complex systems and are difficult for typical operating personnel to understand. There are numerous considerations. The operating screens must be set up to allow the operating parameters to be easily programmed on the screen. Can the operator find all the information rapidly or does it require hunting through all the screens to complete the setup task? The system must be set up to allow sequential changes (delete or add) to the program. Usually, these changes to the relay ladder diagram must be made rapidly or the item being molded cannot be produced. Many times, the original equipment supplier must make these changes to your program. Determine what the turnaround time is to make these changes.

Closed-loop controls are now possible due to the microprocessor. Now, real-time clock functions and math functions are built into these systems which will enable you to produce better parts. Now, SPC controls and reports are easily achieved. Now, the RS 232 ports are standard parts of the blow molder.

A well-designed microprocessor system will include the following:

    A printer access port
    Spare inputs/outputs to expand the system
    Noise-free operation to allow proper functioning of the
        equipment
    A proper ground to the unit
    Control voltage separated from high voltage
    Spare marked wires
    Operator cabinet near press
    Production-oriented screens

The flexibility of the microprocessor has enabled us to get new parts on stream and into production in a very short time. Set-up functions are extremely simple to make and modify with one person. Trouble shooting problems are quickly isolated through the machine sequence screen that minimizes downtime. Now, we are producing parts that could not be produced by the blow molding method five years ago.

2.11 PNEUMATIC SYSTEMS

Air is an extremely important part of producing acceptable parts. The air supply must be constant during the entire production cycle. It seems as if there are never enough air valves supplied with the original machine. The molding sequence must be programmed to handle these added valves. Make sure that the cfm capacity of the main air blow is sufficient to produce the part. Air must rapidly be injected into the parison. It is also important to allow the air to be rapidly discharged (vented) out of the finished part before opening the mold. Following is a list of the types of air functions to be considered for various blow molding applications:

> Low air blow into the initial parison upon mold closing
> Preblow the parison to allow for prepinch mechanisms for needle blow
> Needle blow of main air blow with needle usually located in the mold
> Prepinch of parison to allow needle blow
> Blow cylinder for main air blow
> Stripper mechanism to discharge part out of mold section
> Press gates for automatic operation
> Knock out (cutters) mounted in mold to pierce the finished part
> Spreader to spread (move) the parison; usually used with bottom blow pins that are usually offset on parts
> Pneumatic shut-off slide gate for hopper
> Part removal system for automatic removal of finished parts
> Automatic lubrication system
> Part ejectors mounted inside molds to release parts from molds
> Retractable blow pins

Figure 2.5 SED 320DD Model industrial blow molding system from Davis-Standard, Edison, New Jersey. This unit has two press sections, two heads, and two extruders on a single frame. It shares a common microprocessor and hydraulic unit.

Dual-head operation requires duplicating air values in most cases which doubles the number of valves required. Air storage tanks provided on the machine ensure surge-free air during the operating sequence. This helps produce consistent parts (Figs. 2.5 and 2.6).

Some systems are provided with as many as 32 air valves. Again, make sure that the relay logic diagram (RLD) is properly programmed to handle these functions.

2.12 PART TAKEOUT SYSTEM

Most of the equipment being supplied has dual-head operation. It would be very difficult to have an operator walk into

Figure 2.6 Hartig Model Blow Molder for 55-gallon drums. Includes two 4-1/2 in. extruders, one 35-pound accumulator head, and a two-station shuttle press.

the press area to take the parts out from the platen area. It is also time-consuming to remove parts manually. Automatic part removal provides a more efficient, consistent, and faster cycle time. In fact, using part takeout in an automatic system produces more parts (Figs. 2.7–2.9).

Part takeout is used to remove parts from the mold area provided the mold ejectors remove parts from the mold cavity. If the parts are difficult to remove from the mold cavity, the takeout mechanism will be ineffective in taking parts from the press area. This problem will create lots of downtime.

2.13 SELECTION OF EQUIPMENT

Now, we shall review the overall design of the product (product information).

Figure 2.7 Graham Engineering Model GEC-20DP7440 Blow Mold Unit producing two 32-gallon trash containers on a 60-second cycle. Includes part takeout system. This is a dual head, 20 pounds each unit.

Description	Value
Production requirements	651,000 parts/yr
Shot size each cavity	7 lb
Part weight finished	4.9 lb
Cycle time	70 seconds
Parts per hour/cavity at 100%	51 parts
Parts per hour/cavity at 92%	46.5 parts
Production hours available	7000/yr

After a review of the information estimated and known about the product, we have selected the following blow molding machine to produce the parts. To produce the 651,000 parts per year requires a dual-head machine. We used a yield of 92% and 7000 hours of production time per year. Although an eight pound head is capable of making the parts, we select-

Figure 2.8 Model GEC-22DP7484 Graham Engineering dual 22-pound heads, producing two slides at Hedstrom. Cycle time is approximately 110 seconds.

ed a ten pound head to match the press and extruder size. This also gives more future flexibility to produce a wider range of products.

Output per hour	714 lb
Mold size	25 in. wide by 36 in. high
Mold shut height	30 in. thick
Accumulator head capacity	8 lb (recommended 10 lb head)
Press size	60 in. wide by 40 in. high
Press tonnage	95 tons of clamp force
Extruder size	4 1/2 in. 24:1 L/D rated at 1000 lb/hr HDPE

With these specifications and details well defined, it will make your machinery selection more unbiased. You should

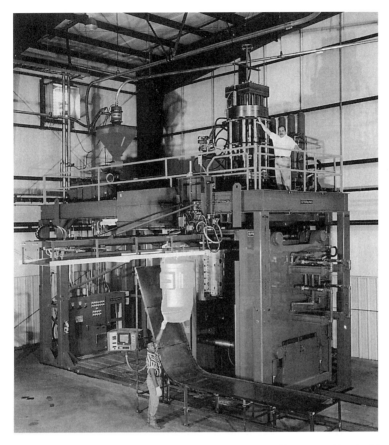

Figure 2.9 Sterling blow molding system model SE675 at Eagle Manufacturing Company, Wellsburg, West Virginia. (From Davis-Standard, Edison, New Jersey.)

make a comparison chart for all the machinery manufacturers being considered. It is best to have at least three competitive quotations. Try to do apples to apples comparison among all suppliers. Some machine producers include a component as standard. Others provide a list of optional subcomponents that may be added to the basic machine price for a true comparison.

It is wise to visit the machine producer's plant to see similar equipment being produced. If at all possible, have the manufacturer run your mold on its equipment in its plant before purchasing any machinery.

You may want to contact by phone or actually visit existing customers of each supplier to get first hand information on the performance and backup support of the machine builder. Good luck and remember to consult with your team of material supplier, mold builder, machinery builder, and designer to fine-tune all necessary project requirements.

3

Start-up and Initial Production

Robert A. Slawska

Proven Technology, Inc.
Belle Mead, New Jersey

3.1 EARLY DAYS

Now that you have selected a new blow molding machine for your project, the fun has just begun.

Now, you must prepare for this new equipment. Many items must be considered and planned for upon arrival of the blow molding system. We have prepared a typical entire blow molding system outline (Fig. 3.1).

It would be wise to have a kickoff meeting with your personnel and the machinery builder. This should be at the machinery builder's plant. It will be helpful to see similar equipment and to have access to drawings and specifications, as required.

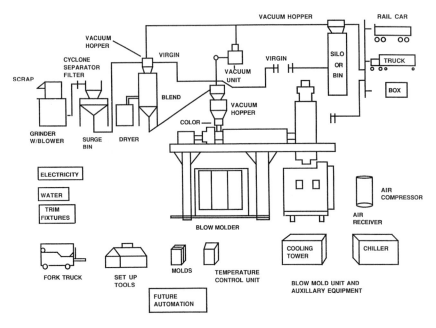

Figure 3.1 Typical blow molding system.

This meeting will review all the specifications of the equipment, such as:

    Machine layout
    Utilities required
    Weight of components
    Operator position
    Location of stairs
    Mold mounting means
    Head tooling size
    Discuss feed screw design
    Location of air valves
    Location of hydraulic valves
    Actual installation

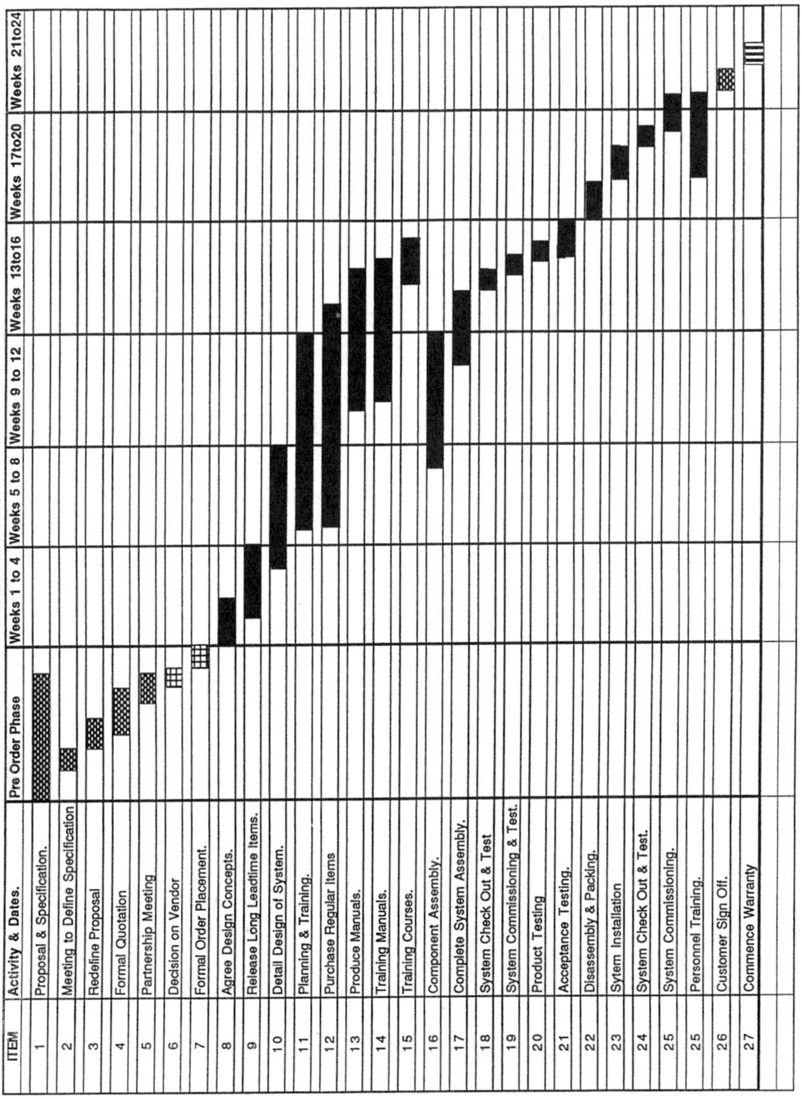

PROJECT TIMESCALES

ITEM	Activity & Dates.	Pre Order Phase	Weeks 1 to 4	Weeks 5 to 8	Weeks 9 to 12	Weeks 13to16	Weeks 17to20	Weeks 21to24
1	Proposal & Specification.							
2	Meeting to Define Specification							
3	Redefine Proposal							
4	Formal Quotation							
5	Partnership Meeting							
6	Decision on Vendor							
7	Formal Order Placement.							
8	Agree Design Concepts.							
9	Release Long Leadtime Items.							
10	Detail Design of System.							
11	Planning & Training.							
12	Purchase Regular Items							
13	Produce Manuals.							
14	Training Manuals.							
15	Training Courses.							
16	Component Assembly.							
17	Complete System Assembly.							
18	System Check Out & Test							
19	System Commissioning & Test.							
20	Product Testing							
21	Acceptance Testing.							
22	Disassembly & Packing.							
23	Sytem Installation							
24	System Check Out & Test.							
25	System Commissioning.							
25	Personnel Training.							
26	Customer Sign Off.							
27	Commence Warranty							

Figure 3.2 Typical activity and dates required to build the blow molding unit.

Project timetables can also be discussed during this initial kickoff meeting. See Fig. 3.2 for typical activity and dates required to build the blow molding unit.

3.2 MACHINE LAYOUT

As early as possible, determine the actual overall size of the blow molding equipment. This will allow you to prepare for rapid installation of the equipment. We have seen plants start up as quickly as one and a half days after arrival from the producer.

Allow proper ceiling heights for proper ventilation, access to service the accumulator heads, and future removal of the heads, if necessary. Your equipment might require retractable air blow mechanisms mounted on top of the head. This could add 2 1/2 to 3 feet to the height of the machine.

Review the plant temperature. Are there any air drafts from doors, windows, air conditioners, or fans? These elements will have a negative effect on the head performance. A cold draft causes the parison to curl or change drop lengths.

3.3 UTILITIES REQUIRED

Information about the utilities required is readily available from your machine builder. The following are the important items required:

Air
How many cfm are needed?
What pressure is available?
How much volume is required?
Is there a pressure drop on the line (is this equipment last on the row)?
Is a surge tank necessary?
What line sizes are required?
Is filtered, dry air needed?

Water
> How many gpm are needed?
> Where are the hookups on the machine?
> What size lines are required?
> What water temperature is needed?
> Is a cooling tower required?
> What size chiller is needed?
> Where is the chiller to be located?

Electricity
> Voltage?
> Breaker size (full-load amps)?
> Location of main breaker?
> Where is the power hookup on the machine?
> Are isolation transformers required?
> Is there enough power in the plant for new equipment?

3.4 WEIGHT OF COMPONENTS

It is absolutely necessary to be advised of the weight of each subcomponent of the blow molding unit. This equipment is large and heavy. You must have the proper equipment to handle each part. It may require a crane, large forklift, or combinations of both of these items.

3.5 OPERATOR POSITION

Once it has been determined where the machine is to be located, you can specify the operator position. This actually is the position on the machine where the blow molded parts are removed.

Space is required to finish the parts by removing the flash (trim) and doing any drilling or cutting required. Room must also be provided for auxiliary equipment, such as grinders, conveyors, and pack out boxes.

3.6   LOCATION OF STAIRS

Stairs are provided for access to the upper level of the equipment. It is important to locate these stairs so that access is safe and so that they do not interfere with production functions. These stairs must be built to OSHA and ANSI safety specifications.

3.7   MOLD MOUNTING MEANS

In considering the machine layout, it is important to identify the means of mounting your molds. Will you require a forklift truck, overhead crane with a roll out clamp, or some other means of lifting and mounting molds? Whatever your choice, space must be provided to accomplish mold changeovers. This may require moving worktables and conveyors.

When the equipment is ordered, you should also determine how to attach molds to the platens. The two most common means are bolt holes or "T" slots. Locating slots, pins, or bars may also be used for mounting molds.

Usually the actual mounting of the mold is accomplished within an hour. The time consumed is hooking up all of the cooling water lines, air connections, hydraulic mechanism attachments, and blow mechanisms. Quick connection fittings, manifolds for air, water, and hydraulic connections will drastically shorten the changeover time.

3.8   HEAD TOOLING SIZE

Once you change molds, is it necessary to replace the head tooling with a new size? Most likely, this will be required. The new size must be determined by experience. Like mold changeover procedures, you must determine the best way to change head tooling. If the press is rolled out of the way, head tooling can be changed at the same time as the mold is being replaced. Remember that the head tooling is heavy and very hot. You must be set up properly to achieve this switchover.

3.9   FEEDSCREW DESIGN

Determine the material that will be run in this equipment. This should be reviewed before it is completed. For your application, a general purpose feedscrew might be required. This will allow producing several different polyolefins in the extruder with minor disadvantages. However, if you also plan on running engineering resins, a new screw might be required. Remember, if a feedscrew changeover is necessary, provide the space to accomplish the switchover.

3.10   LOCATION OF AIR VALVES

It is best to locate the main air valves as close as possible to the inlet of the parison. This assures very rapid inflation of the parison. The air lines and the valve and manifold should be as large as possible, at least 1 in. in diameter. For very large parts, a 2 1/2 in. diameter should be considered. Likewise, large lines are beneficial for quick exhaust of the internal air within the part during the vent cycle. The exhaust valve must be located right at the discharge of the blow needle or blow pin. Some parts require low blow air with low pressure. This should be connected to the main air blow lines, where required, to allow rapid transfer between the two. Preblow air most be set up to be part of main blow air or separate when preblow is in a location different from the main air blow. In both cases mentioned, if the main air blow is directly connected to preblow or low blow, check valves must be used in the connecting lines. This will prevent the main air from leaking out through these auxiliary valves. In most cases, the other air valves used in blow molding are 3/8 in. This includes the gates, needles, knockouts, prepinch, and strippers. Most times these valves are manifold mounted either on the press or on the main frame.

3.11   LOCATION OF HYDRAULIC VALVES

Like the air valves, it is best to locate the hydraulic valves near the actual function they control. Valves for split molds, knockouts, and unspin motor devices for threaded blow pins; parison spreaders; and up/down movement of blow pins are usually mounted on the press frame. For most of these of the previously mentioned functions, the valves should be mounted in the rear center of the platens (head centerline).

The shoot and program valves should be mounted near the accumulator heads on top of the main base. The open/close and high tonnage lockup valves should be mounted on the press. Again, either the valves or the piping to the cylinders should be centered.

Care must be taken with the actual valve mountings. Remember that hydraulic valve and line connections leak due to high pressures and shock. It is recommended that drip pans be mounted under the valves to capture small amounts of leaking oil. Drip pans also allow O-rings to be changed, and the oil leakage is caught by the drip pan.

3.12   CUSTOMER ACCEPTANCE/PRESHIPMENT TESTING

The machinery supplier has just called you and lets you know that your blow molding machine has been completed. You will want to send your molds and material to the machinery builder for testing. Once this testing is in process, you will want to perform a series of checkouts before releasing the machine for shipment.
Review the following items for your checkout list:

Perform a visual check of the machine.
Look for loose or leaking hydraulic lines or connections to valves.

Check all temperature zones for set points versus actual temperatures.

Perform an extruder output check.

Confirm that the head calibration is working and set up properly for programming and shoot out.

Set operating hydraulic pressures and oil flows required.

Insure that the hydraulic oil temperature is running in a proper range (110 to 120°F).

Reinspect the hydraulic system for leakage once the machine has functioned for at least five hours in normal operating ranges.

Verify that the head tooling is mounted and calibrated for proper mechanical zero and electrical position.

Inspect safeties for correct number, location, and performance.

Check the LVDT mounting brackets and tolerances.

Run the machine in a simulated cycle before using plastic in the system.

Calibrate the press plate zero position with molds.

Enter the necessary production set points for microprocessor start-up.

Insure that the full sequence of operations is correct and operating to your expectations.

Verify press clear platen area, length, and width.

Verify minimum and maximum press daylight sizes.

Verify clamp tonnage.

Confirm shot capacity of the head with the extruder off.

Confirm extruder size and motor H.P.

Determine hydraulic unit H.P. capacity.

Determine pump capacities of the hydraulic unit.

Verify platen open/close speeds in various ranges.

Check shooting speeds of parison drop rates.

Test all vertical and horizontal adjustments of the press.

Verify the melt temperature of the parison.

Test and verify that the parison programming steps are functioning.

Verify that hydraulic schematics are correct.

Review electrical wiring diagrams.
Verify that the barrel cooling system is working.
Review the technical service manuals.
Verify that the optional equipment is complete and oper-
ating.
Verify that the melt pressure gauge is functioning.
Verify rupture disc protection.

If you accept the machine for shipment, note any corrections or missing items that need to be completed. A timetable should be established and agreed upon before shipment as to whether the item is completed before shipment or at your plant.

It might be desirable to have both formal and hands-on training sessions during the preshipment trials. All aspects of the equipment should be covered during this training, including the sequence of operation. Your personnel must be provided with in-depth knowledge of the entire blow molding unit. This will include a general machine layout and mechanical features. Pay special attention to the electrical systems, microprocessor controls, and hydraulic components of the machine. These areas are generally problematic for molders and set-up personnel. You will find out that many of the machine controls are read as a number in percent of "something." You must understand what that "something" is.

After the equipment has been approved for shipment, a shipping date must be determined. The appropriate number and type of trucks can be ordered for transportation to your plant. Don't forget to insure the equipment during shipment with a sufficient amount of coverage. At this time, the equipment is disassembled and painted to prepare it for shipment. This could take up to seven days.

Arrival of the equipment must be scheduled with the rigger. Because rigging costs are high, the timing of the delivery must be coordinated closely.

Upon completing the reassembly in your plant, the machine manufacturer should be contacted. It is wise to provide as much advance notice to allow the scheduling a technical service person to start up your equipment. Usually the builder will call you to verify that all utilities are hooked up to the machine and that the machine is ready for start-up.

3.13   READY FOR START-UP

Once the service person arrives in your plant, a complete check of your equipment similar to the preshipment testing and as shown below is performed:

Verify if any parts are missing or damaged in shipment.
Is the machine ground properly installed?
Verify incoming voltage on all three phases.
Visually check the machine to assure proper installation:
Loose electrical connections?
Retighten heater bands and straps.
Retighten/check all bolt connections at proper temperature.
Proper motor rotations.
Customer's water connection to heat exchanger.
Air connection.
Verify that hydraulic oil is filtered to 5 microns absolute.
Verify that head and machine are level.
Check for oil and water leaks.
Verify that molds are installed properly.
Calibrate the clamp zero point with molds mounted.
Ensure that all hydraulic hose are in proper location and tightened.
Calibrate tooling (zero and span).
Set mechanical tooling stops.
Check for alarm conditions on operator screens.
Check operation of melt pressure gage.
Check operation of rupture disc limit switch.
Modify sequence of operation as required.

Upon successful start-up and checkout, now you will begin the production cycle on your new equipment. We all wish that things go well and that the equipment runs correctly. However, from time to time new components will fail during early stages and must be replaced.

Your training and knowledge of the equipment can quickly pinpoint the problem and most likely solve it in a short time. Being prepared to face these failures with proper spare parts and replacement timing knowledge will minimize your downtime.

3.14 SUMMARY

Your goal is to achieve optimum production in the shortest time. This is possible based on the following considerations:

Early planning
Provide all information quickly to the machine builder.
Establish what is being built versus what you want.
Prepare your plant for the new equipment.
Inform your rigger of all details.
Get as much training as possible.
Do thorough preshipment testing.
Be prepared to receive all of your equipment.
Plan and be prepared for installation.
Be ready for the service engineer and start-up.
Start production.

4

SELECTING A BLOW MOLDING MATERIAL

W. Bruce Thompson

Fluoroware, Inc.
Chaska, Minnesota

Blow molding is a very complex process because a large number of properties are balanced and interdependent. It seems simple to melt some plastic, make a preform shape, close a mold around it, and inflate a hollow object, but consistency is the whole key to blow molding at production speeds. Achieving consistency is a matter of paying attention to all of the details involved in running the blow molding machine.

In general, blow molding requires that plastic has high stiffness, good impact properties, good environmental stress cracking resistance (ESCR), and processes consistently.

Most plastic parts require specific physical performance properties that provide a benefit over an alternate material. For example, a lightweight bottle is desirable, so that a lot of the freight cost goes into the product and not into the package. Contrast the very light weight of a one-gallon milk bottle with the heavier weight of the glass bottle that was replaced by plastic a few years ago. The density of plastic is much lower than glass, but the wall thickness of the plastic milk bottle is much thinner than the glass bottle because the plastic bottle has better performance properties at lower thickness. The performance property that allows the bottle to have such a thin wall is flex modulus, a measure of stiffness. The glass bottle far outperforms the plastic bottle in stiffness, but maybe the plastic bottle is good enough for the application so the flex modulus value doesn't need to be as high as that of glass. Similarly, the plastic bottle also has good impact properties at the thin weight whereas the glass bottle might be considered dangerous or brittle if the wall thickness is too thin.

It is also very important to know that one property isn't always the only factor for physical performance. In the milk bottle example, density also contributes to the stiffness of the plastic, because of the type of polymer. Homopolymer polyethylene has a density >0.960 g/cc whereas copolymer polyethylene often has a density of 0.950-0.955 g/cc, and homopolymer polyethylene is much stiffer than the copolymer because of the carbon chain structure. The homopolymer is linear and has a higher degree of crystallinity. The copolymer is branched, and the carbon chains cannot pack together as tightly because of the space occupied by the branched molecules. Both homopolymer and copolymer polyethylene could make a milk bottle, but the homopolymer outperforms the copolymer because the bottle can be made slightly thinner, therefore at lower cost per package.

4.1 MATERIAL SELECTION PROCESS

When selecting resin for a part similar to an existing part, the obvious choice is to start with the current resin, if known. Sometimes it is very difficult to tell the difference between similar resins. For example, acrylic, polycarbonate, and polystyrene are all clear rigid materials. Polyethylene and poly-propylene are opaque, rubbery materials, but usually it is easy to distinguish between dissimilar resins, such as polystyrene and polyethylene. At some point, it is prudent to make a part from more than one material and then conduct performance testing to determine how closely each resin meets the requirements or which resin gives the best performance combination. Cost should be the last consideration in the material selection process (Fig. 4.1), a part that delivers value, because it is closely matched with the needed performance characteristics, is much better from your customer's perspective.

There are many ways to find the resin that you want to use. Manufacturers have data sheets that list performance

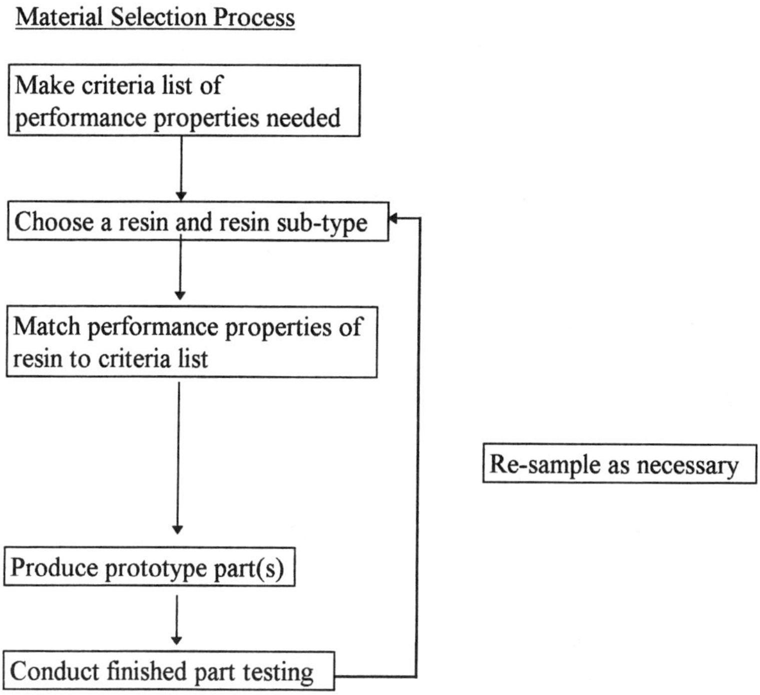

Material Selection Process

Figure 4.1 Material selection process.

characteristics of the resin, and there are many computer list-
ings available that show resin differences between suppliers.
Data sheets show what the manufacturer wants to highlight
for the resin. There is no consensus on the data that needs to
be reported. Most often data sheets for blow molding resins
show melt index, density, tensile, elongation, and modulus.
Sometimes other properties are shown, and this can be a good
clue to what the manufacturer suggests are the benefits of
that resin.

Some resins are more notable because of their specific
properties. For example, polypropylene has a higher heat distortion temperature than polyethylene and is used for hot-fill
applications that may distort a polyethylene package. Polyethylene is commonly used for moisture-containing applications because it is rather impervious to water. Many household package applications are aqueous and benefit from
packaging in polyethylene. Polyethylene is also rather impervious to many harsh chemicals and solvents and therefore is
used to package them. Some of the light hydrocarbons are not
good choices for packaging in polyethylene because the porosity of the resin is actually high and the hydrocarbons may
percolate through. Polyethylene can be fluorinated or sulpho-
nated to make it less permeable to hydrocarbons. Sometimes
the resin makes a difference in the materials it can package.
For instance, polyvinyl chloride is a good choice for packaging
a product that contains a light aromatic solvent because it is a
fairly nonporous material.

Polyethylene is not a good barrier to oxygen, but it has an
adequate service life in packaging milk. Polyethylene teraph-
talate (PET) is a good choice for packaging products containing light gases (O_2, CO, and CO_2), whereas the olefins (polyethylene and polypropylene) do not hold those gases very
long.

Many choices face the package designer, not the least of
which is resin choice. Obtain as much information from litera-
ture and product data sheets as possible. Then, undertake a
rigorous testing program to determine the optimum package.

4.2   MELT INDEX

The melt index is a measurement that indicates the flow of material under known conditions. Often referred to by different names, (melt flow rate, melt viscosity), melt index provides a relative measure of the flow of material in the molten state (Fig. 4.2). The test is always performed according to strict guidelines that give consistency from one test to the next or from one test facility to another.

The melt index test uses a hollow cylindrical die (0.0825 inch opening) to form an extrudate. The thermoplastic test material is placed in a heated chamber, brought to the test temperature, and then extruded from the die by the force of a weight acting on the plunger in the heated chamber. A test that dictates the conditions for conducting melt index tests is ASTM D-1238. There are several conditions within this test that correspond to those needed for various materials. Condition 190/2.16 g/10min is common for general purpose polyethylene used in bottle making up to about a four-liter size. The temperature in this condition is 190°C, the weight used is

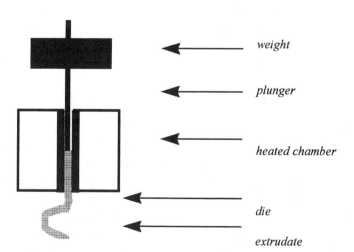

weight

plunger

heated chamber

die

extrudate

Figure 4.2   Diagram of a melt index test machine.

2.16 kg, and the g/10 min is the weight (in grams) of extrudate measured in a ten-minute test. Other materials use different weights and conditions. For example, polypropylene is tested at 220°C. Higher molecular weight polymers require much larger weights because there wouldn't be much flow with those materials at the 2.16 kg condition. Often called the HLMI (high load melt index), the weight used is 21.6 kg for high molecular weight materials. Sometimes the melt index test is performed at more than one condition to produce a ratio (example, I_{10}/I_2) which gives some indication of the flow at two extremes.

Melt index is useful as a quality control reference number. Request melt index test results from your supplier, with each shipment of product as a quick reference that the received material is as ordered. Keeping track of the numbers will give you some indication of the consistency of the resin.

4.3 DENSITY

Density is weight per unit volume, usually expressed as grams per cubic centimeter but sometimes as pounds per cubic foot. Density provides some indication of how closely the molecular chains of similar materials are packed. Between product types, for example, polystyrene and polyethylene, density is not a useful measure except to say that polystyrene is heavier than polyethylene. Between homopolymer and copolymer polyethylene, density indicates the amount of co-monomer in the resin.

Density is measured by different methods. ASTM D-1505 requires a density column which is a graduated cylinder about three inches in diameter by about four feet long filled with carefully controlled liquids of varying density (Fig. 4.3). The heaviest liquids settle to lower levels, and the light liquids float to the top. Beads with known densities are placed in the column and float at the level matching the liquid density. A test pellet is dropped into the column and allowed to settle

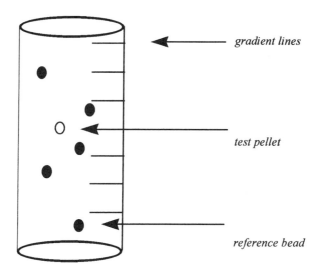

gradient lines

test pellet

reference bead

Figure 4.3   Density gradient column.

to the level where it floats. That corresponds to the density of the liquid medium. Then, the density of the pellet is read from the graduated scale on the glass column.

Bulk density is the weight per unit volume of the plastic in pellet form, sometimes derived by filling a known cubic measure with pellets and weighing the net weight. This is also a useful number for the blow molder because it allows sizing the material handling equipment, such as blenders, loaders, and silos.

It is very important to eliminate all sources of material contamination from the handling system. When designing a material handling system for blow molding, ensure that long radius corners are used in all material lines and that pellet velocity is kept at a medium level. A phenomenon known as angel hair, the smearing of the pellet out into a long fiber, is caused when pellets move too fast through piping. The fibers collect in hopper loaders and blenders and slow down or stop the flow of pellets to the machine.

4.4 STRENGTH PROPERTIES

4.4.1 Tensile

Tensile strength is the extensible resistance to breakage of a material expressed in force per cross-sectional area. A load frame is used to provide a constant force on the sample in a straight line direction until the sample breaks. Because the sample has to be restrained in the sample holder, the force of the clamps squeeze the sample with small teeth or knurling, and the sample would break at this point from the reduced cross-sectional area. Instead the test sample has a dog bone shape (Fig. 4.4) to concentrate the test force in the center region of the specimen and provide larger clamping regions on each end.

There are many shapes used for tensile testing. ASTM D-638 defines many of the conditions of tensile testing, including which coupon shape produces results as described in the precision and bias data. The material being tested and its extension curve has a lot to do with the shape of the coupon, how fast the extension force is applied, and how fast the coupon breaks (Fig. 4.5).

Tensile properties are very useful to the part designer. Mechanical forces within the geometric shape greatly influence the radius of corners and the thickness of the part. It is also useful to audit tensile properties periodically to verify that your part has the same minimum values with which it was designed.

4.4.2 Elongation

Elongation is a measure of a material's ability to flow under force in a given time period. The elongation test is normally

Figure 4.4 Typical dog-bone shape of a tensile test coupon.

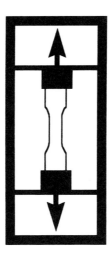

Figure 4.5   Diagram of a load frame. The clamps hold the test coupon rigidly in place as the machine applies mechanical force in an extension direction. Many load frames use powered lead screws to generate the force and control the speed of extension.

conducted concurrently with the tensile test. A material described as rigid, brittle, nonductile, or any combination of these, would not elongate very much before breaking, whereas a material that is described as soft, pliable, or rubbery, might elongate a great amount before breaking. The reason for the amount of elongation before breakage can be traced to the morphology of the material. Plastics that have straight molecular chains, very few side branches or bulky elements, or chains that are long and have many folds (lamellae) elongate fairly easily. Plastics that have many branch chains, irregular structure, or cross-link ties between chains, don't elongate very much.

The elongation property is not the same as tensile strength. It is only the amount, relative to the original size, that the material changes under extension forces. Relative measures of elongation are also important because they describe the elongation up to a certain point. For example, elastic recovery describes the amount of elongation before permanent change is introduced into the material.

4.4.3 Tensile Yield

Tensile yield is another property measured during the tensile test. It is defined as the lowest stress (material's response to applied force) under which a material shows plastic deformation. Below this stress value, the material exhibits elastic recovery and would return to its original state (Fig. 4.6).

4.4.4 Flexural Modulus

Flexural modulus is defined as a material's resistance to bending. In many thermoplastic materials, the response of the solid material (stress) to an applied strain (force) is proportional through a factor called the modulus. This relationship in a solid material is defined as Hooke's law. The modulus is a measure of the material's stiffness or ability to resist deformation. ASTM D-790 is a commonly used test for measuring flexural modulus (Fig. 4.7). Normally conducted in a load frame, the extension force required to bend the sample coupon

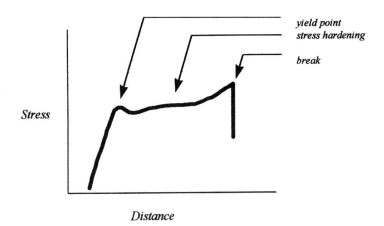

Figure 4.6 Material response in tensile conditions.

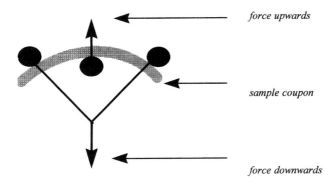

force upwards

sample coupon

force downwards

Figure 4.7 Flexural modulus test.

a known distance is measured and expressed as flexural modulus.

Filler compounds are used to enhance the performance of thermoplastics and/or lower the cost by extending the material. Fillers are reinforcing if they have a somewhat cylindrical shape. Aspect ratio is the relationship of the length of the mineral or fiber to its diameter. Higher aspect ratios provide great increases in flex modulus values. Filler materials that have undefined or very low aspect ratios act as extenders of the thermoplastic. An example of an undefined aspect ratio is a glass microsphere. An example of a high aspect ratio filler is a glass fiber about 0.5 in. long.

4.5   IMPACT PROPERTIES

4.5.1   Drop Impact

Drop impact testing is performed on the final product under conditions relevant to the part's normal usage and storage. The part or package (and contents) is conditioned to one of three temperature ranges, room temperature, very low temperature (often –40°C), or moderately high temperature (often 50°C). Most testing consists of dropping the package from a repeatable level and inspecting for damage. Statistics can be employed to predict part performance. Low temperature impact testing for packaging, often conducted at –40°C, requires that a liquid remain as a liquid at that temperature. Liquids produce a hammer effect within a container that is not present in solid shapes and concentrate the impact force into a smaller area than the large area associated with the solid. Failures of the package occur first at the weakest point (defect).

4.5.2 Izod Impact

Izod impact is a measure of a material's impact resistance. A test specimen that has a notch to initiate crack propagation is placed in a holder, and a weighted force swings through an arc and breaks the specimen (Fig. 4.8). The force is measured as foot pounds per inch or as joules per meter.

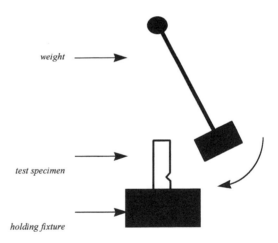

Figure 4.8   Izod impact test.

4.5.3   Environmental Stress Crack Resistance (ESCR)

ESCR is the material's resistance to developing cracks (or crazing) due to the influence of chemicals and stress. ASTM D-1693 is a test for measuring the number of hours of exposure to a chemical stress where 50% of the population of test specimens fails (Fig. 4.9). Sample specimens are cut from bottles or compression molded plaques and placed in a fixture that holds them in a bent or U-shape while immersed in a detergent solution at elevated temperature. The test is stopped when the coupons fail by cracking.

4.5.4   Outdoor Exposure

There are many factors relevant to the performance of a plastic part exposed outdoors. Ultraviolet light from sunlight is a high energy source that generates free radicals in a polymer. UV light stabilizers added to the resin inhibit their formation by acting as UV specific energy absorbers or as radical scavengers. Hindered amine light stabilizers (HALS) are very common UV additives. The other common form is a light

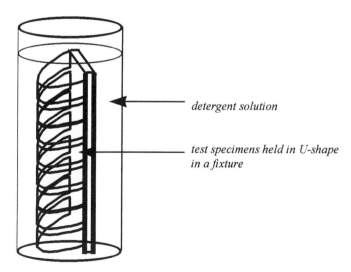

detergent solution

test specimens held in U-shape in a fixture

Figure 4.9   Environmental stress crack resistance.

blocker which prohibits the UV wavelength energy from reaching very much of the plastic.

Thermal stability is very important to the blow molding process because one of the common effects of heat without adequate stabilizer concentrations is chain scission of the plastic molecule. The net effect of chain scission is the same as lower molecular weight, that is, lower melt strength, lower impact strength, and lower tensile properties. Oxygen readily available from the atmosphere also induces the long chain carbon molecule to break into shorter lengths. Other environmental factors include active metals and aggressive chemicals.

Thermoplastics for blow molding are designed with about enough stabilizers to cover five to seven extrusion cycles, depending on the conditions of extrusion and the type of polymer. Extrusion blow molding generates flash (roughly 35% by weight) but one of the benefits of thermoplastics is that the flash can be reground, mixed into the feed stream, and reused. It is important to limit the endless loop of the flash regrind system so that all of the material is consumed. A periodic total cleaning out of the regrind system ensures the stabilizer system is not being consumed in just protecting the resin during extrusion.

Choice of the finished part's color is also important. Carbon black, the main ingredient in black color concentrates, is by itself a very good light stabilizer. White products don't absorb as much heat as darker colors. Most color concentrates are formulated for the specific job they are intended to do. An all-inclusive color additive can be very expensive, but again value should be present if the material meets the part's requirements.

Antistatic additives are another additive that is very important to the packaging industry. Low molecular weight polymer waxes are sometimes used to dispel static charges by blooming to the surface, attracting moisture, and dissipating the charge. Blooming antistatic additives are effective for the long term but distinctly affect printing on the container. Another treatment for static electricity involves using metal or conductive fibers touching the parison as it is being formed and pulling the electricity to ground. This treatment is effective for the short term, i.e., package formation.

4.6   MELT PROPERTIES

Some terms are used with plastics processing that deserve some explanation. Rheology is the study of the deformation and flow of material in terms of stress, strain, and time. Stress is the distribution of force over an area of the material, strain is the change in the material, and changing strain with time is the shear rate (Fig. 4.10).

In an ideal solid, Hooke's law describes the behavior of the solid when a constant stress deforms the material immediately to a constant strain, but it recovers totally when the stress is completely removed. Examples might be a coiled metal spring or a good rubber band. However, in an ideal fluid, the opposite is true. Newton's law describes a material's response to a constant stress that deforms it immediately, but when the stress is removed there is no recovery. An example of an ideal fluid is water.

For a non-Newtonian fluid, viscosity is defined as

$$\eta \equiv \frac{\sigma}{Y}$$

4.7   MATERIAL RESPONSE TO FORCE

A viscoelastic material combines both Hooke's and Newton's principles to exhibit time response but incomplete recovery. Creep is a term used to describe the change in a material due to stress over time. The zero-shear viscosity (lowest point on the shear stress curve), which is a creep measurement of a polymer melt, is an approximate measurement of the melt strength of the material. Creep measurement on solids pro-

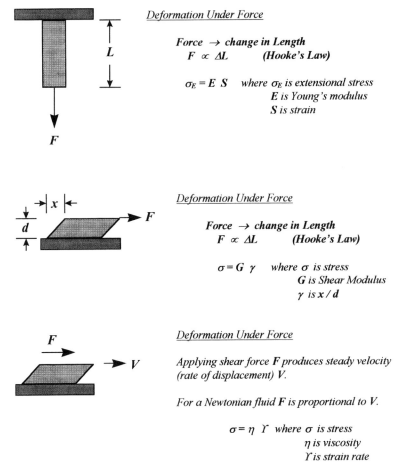

Deformation Under Force

> ***Force → change in Length***
>   ***F ∝ ΔL***   *(Hooke's Law)*

$$\sigma_E = E\ S \quad \text{where } \sigma_E \text{ is extensional stress}$$
$$E \text{ is Young's modulus}$$
$$S \text{ is strain}$$

Deformation Under Force

> ***Force → change in Length***
>   ***F ∝ ΔL***   *(Hooke's Law)*

$$\sigma = G\ \gamma \quad \text{where } \sigma \text{ is stress}$$
$$G \text{ is Shear Modulus}$$
$$\gamma \text{ is } x\,/\,d$$

Deformation Under Force

*Applying shear force **F** produces steady velocity (rate of displacement) **V**.*

*For a Newtonian fluid **F** is proportional to V.*

$$\sigma = \eta\ \Upsilon \quad \text{where } \sigma \text{ is stress}$$
$$\eta \text{ is viscosity}$$
$$\Upsilon \text{ is strain rate}$$

Figure 4.10   Examples of deformation under force.

vide design information for load bearing capabilities, deformation, and aging properties (Fig. 4.11).

Many plastic materials recover if the deformation distance or deformation rate is not too large. Some materials exhibit a memory if allowed long enough to recover, and they may return almost to original shape.

Figure 4.12 shows the yield stress of the material response to force.

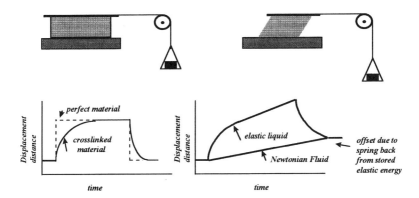

Figure 4.11   Material response to force: creep and recoil.

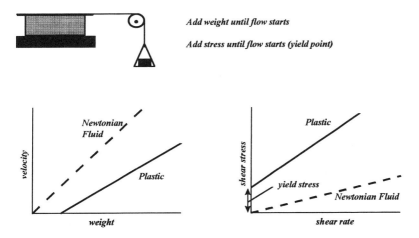

Figure 4.12   Material response to force: yield stress.

Torque rheometers are used to generate viscosity data. Many more than three types are used but the concentric cylinder, cone and plate, and double plate are most common (Fig. 4.13). Examples of problems that can be solved by using rheometers are melt fracture, processing differences between "good" and "bad" lots, improving impact performance (by a

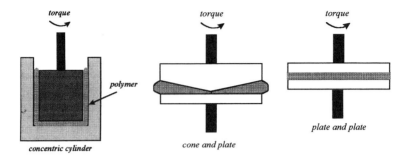

Figure 4.13 Types of torque rheometers.

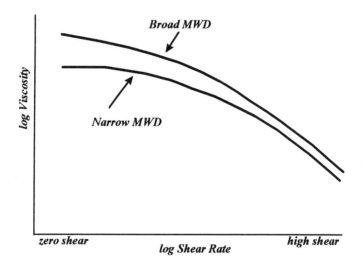

Figure 4.14 Effect of molecular weight: viscosity versus shear rate.

combination of drop-testing and rheometer measurement), and molecular weight differences in polymer melts (Fig. 4.14).

4.8 CONCLUSION

When selecting a material for blow molding, use all of the information available to differentiate between material types and performance properties (Table 4.1). Not very many blow

Table 4.1

Change of Polyethylene Properties with Density, Melt Index, and Molecular Weight Distribution

Property	As density increases	As M.I. increases	As M.W.D. increases
Abrasion resistance	↑	↓	
Barrier properties	↑		
Blocking resistance	↓	↓	
Brittleness resistance	↓	↓	↓
Brittleness temperature	↑	↑	↓
Chemical resistance	↑	↓	
Cold flow resistance	↑	↓	↑
Dielectric constant	↑ (slight)		
ESCR	↓	↓	↑
Gloss	↑	↑	↓
Hardness	↑	↓ (slight)	
Haze	↓	↓	
Impact strength	↓	↓	↓
Load bearing props (long term)	↑	↓	↑
Melt elasticity		↓	↑
Melt extensibility		↑	↓
Melt strength		↓	↑
Melt viscosity		↓	
Mold shrinkage	↓	↓	
Permeability	↓	↑ (slight)	
Refractive index	↑ (slight)		
Shear stress (critical)		↑	↓
Softening point	↑		↑
Specific heat	↓ (slight)		
Stiffness	↑	↓ (slight)	
Surface roughness		↓	↓
Tear strength (oriented film)	↓	↓	
Tensile elongation (at break)	↓	↓	
Tensile modulus	↑	↓ (slight)	
Tensile strength (at break)	↑	↓	
Tensile yield stress	↑	↓ (slight)	
Thermal conductivity	↑		
Thermal expansion	↓		
Transparency	↓	↓	
Viscosity, rate of change with shear rate			↑

Table 4.1 Continued

	Decrease in average molecular weight	Increase in M.W.D.	Increase in temperature of measurement
Melt index	↑		↑
Impact resistance	↓	↓	↑
Chemical resistance	↓		↓
Density			↓
ESCR	↓	↑	↓
Tendency of polymer melt to melt Fracture	↓	↓	↓

molding companies have the equipment available to do all of the testing mentioned in this chapter, but results of these tests will enhance your knowledge and your quest for consistent processing. Your material supplier often has the resources to do the testing; your challenge is in proving that you need the information or finding a willing alternate source.

5

Extrusion Blow Mold Design

**Charles Carey, Tar Tsau, Reinhold Nitsche,
and Karl Schwarze**

*Wentworth Mould and Die Co. Ltd.
Hamilton, Ontario, Canada*

5.1   SELECTING A MOLD MAKER/MOLD DESIGNER

5.1.1   Experience

The selection process that the blow molders go through in deciding what mold maker to select must be a very arduous process. We say this from experience, being front and center in many technical meetings and posed with the following statement, "It is nice to finally meet the person responsible for all the blow molding problems." Certainly this is intended as a "tongue in cheek" statement, but the validity is that the mold, more so the cavity of the mold, is the last thing we see after the container is produced. "The selection of the mold maker is one of the most important decisions the processor can make."

In selecting a mold maker the most common talked about buzz words are quality, price, and delivery. One that is sometimes forgotten up front is service, probably the one that makes or breaks any project at any point in the schedule. The impact that experience plays in the whole process is second to none. Historically, you can background check your mold maker's background through your respective plastic associations, industry consultants, and even competitors to better understand your prospective mold maker. Being old does not necessarily make them the best but certainly puts them in a position whereby many uncharted obstacles have been overcome in the past.

An experience that comes to mind was a difficult and challenging decision for us to mold an off-parting line handle in a three-gallon polycarbonate water container. Keeping in mind that this "challenge" was taken in the mid 1980s, we felt it would be a very difficult, if not near impossible, request. As we all know, once we move away from the natural parting line with a handle 90° to the parting line, we venture into the "it can't be done" territory. Among the many things we were concerned about were the following:

1. The negative effect on cycle time.
2. How do we insert the premanufactured handle into the blow mold?
3. Will the plastic blow sufficiently around the handle itself?
4. And last but certainly not least, After blowing, will the handle support the approximately 30 lb of weight when carried?

To make a long story short, "Murphy's law" reigned supreme as it usually did on this project, and our concerns were all completely laid to rest except for one area. As mentioned, the handle was premanufactured and in fact was blow molded. The handle had small nubs on the end for welding off onto the container and was needle blown. One of our main areas of concern was blowing around the handle itself and, of course, holding the weight. Well, the problems as we saw them never materialized but the plastic blew into the needle opening of the handle allowing water into the handle itself. From a sanitary standpoint, this is obviously totally unacceptable would have to be overcome. The end result was that the handle was injection molded, eliminating the needle hole and allowing the blow molder to proceed with production.

Suffice it to say, this project was a learning experience and certainly falls in the vein of true research and development. We have used this experience many times over in working on related projects that never would or could have been foreseen otherwise. In working to a successful completion, we had to collaborate with the designer of the container, the resin suppliers, a consultant and of course the blow molder. Projects such as these require a true team approach and ownership by all parties involved to "make the thing work."

5.1.2 Reputation/Reference

It has always been our approach to build strong working relationships with our customers' partners. These partners include but are not limited to the resin suppliers, the ancillary equipment suppliers and, from our standpoint, first and foremost the machine supplier. We have a machine database library that is second to none. With support from the blow molder, the mold maker should be prepared to assume the responsibility for the complete design of the blow mold and how it relates to the machine. We have been able to build this database by working closely with the machine suppliers, treating them with the same respect and passions as we do our customers. The mold maker needs the machinery supplier and they need us. Initially, some machinery suppliers are skeptical of the mold maker's intentions. The mold makers should look upon themselves as "free agents," putting the customer and the molds themselves in the forefront. Our motto is that our customer wants to put the best mold on the best machine. There is only one way we can do this and that is by working together.

5.1.3 Delivery

I don't know who first said it, but someone once said, "Build a better mousetrap, and the world will beat a path to your door." The 1990s have brought about a revolution in a "fast to market" approach to deliveries. Mold makers have to realize that you have to be on time and also you have to be faster than ever before. The motto just mentioned in a modified form will come to pass. The mold maker's organization must be set up so that there is a passion for deliveries. It must be felt through the entire organization. If there is a breakdown in the chain, then somewhere sometime it collapses. In our plant team members are encouraged to act quickly and independently. This philosophy must be supported by management by simplifying the design process. Fewer administrative layers and improved communication is the groundwork that must be adhered to for this passion of delivering on time and fast.

All of this can be accomplished, but it has to start from the top. You can't expect to walk onto the shop floor and simply tell your toolmakers or mold designers to work twice as fast. You have to provide them with the tools to do so. A mold maker who has in place dedicated "fast to market" mold teams, "stock block" mold programs, internal incentives to deliver on time, and the competitive passion for never missing a date, will in the vast majority of cases, be the mold maker of choice.

5.1.4 Value

When selecting a mold maker purely from a price standpoint, the decision is easy. As a global mold supplier, we are always confronted with the price issue and when trading outside of

North America, it can't always be possible to find somebody cheaper. This is not sour grapes. Without a doubt, price is an underlying ingredient and one of the common denominators in assessing the competition, but it should not be the determining factor.

The mold maker should always be addressing ways and means of becoming a low-cost manufacturer. Depending on the size and overheads of the mold maker, producing the lowest cost possible should always be a focus. Once again, the "work twice as fast" philosophy will not work. We have to provide them with the tools. Technology plays a pivotal role in making this all possible. Any mold shop worth its salt should be fully integrated with a high-end 3-D CAD/CAM/CAE system.

All related metal cutting machinery should be networked for downloading programs directly to the machines. High speed machinery is a must. Conventional machining of the mechanical aspect of the molds, the waterlines, venting systems, and machine mounting linkage, should always be done whenever possible in one setup. All machinery should be equipped with multiple pallets to maximize efficiencies, therefore, reducing costs and delivery time. Multiple setups add cost to the molds and also have an obvious adverse effect on delivery timing. High speed machining of the cavity is one area where the biggest gains have been made in recent years. In the past, somewhat conventional methods were used that were labor-intensive and left a finish in the cavity that involved a tremendous amount of hands-on time that was difficult to control. Machining on horizontal machining equipment with exceptionally high spindle speeds all but eliminates the old benching technique and this again tackles, head on, the price and delivery consideration.

In competitive markets, the mold maker must be prepared with an organization that thrives on change. Simplicity is the key and should shape the way they do business. Simplicity in workflow, organization, and task reduction, all amounts to less bureaucracy, better designs, improved methods, lower costs, and better value for our customers, too.

5.1.5 Location

In today's global market the idea of location, always being close to your customer, is a very objective and personal decision on the part of the blow molder. It is the industry opinion that in any organization if sustained growth is your mandate, then you have to be willing to set up shop in that geographical sector. From experience, the customer does not want to talk to an agent, rather, they want to converse with a fully qualified sales engineer from the mold company itself.

The mold maker dealing in the global market though must be prepared to provide the service that a mold shop "next door" could provide. The mold makers' philosophy should be to act first and ask questions later. For example, we experienced an incident in shipping to the southeast Asian market where an unknowing customs officer tried to open up molds for visual inspection. Upon arrival in our customer's plant, O-rings had come loose, and the molds were spraying like a lawn sprinkler. For a day and a half we tried to explain and solve the problem by phone but to no avail. We were not thinking or acting globally. We sent a service technician on the next available flight, and it was repaired the following day. This was a hard lesson, but one that has to be taken for granted when, as a mold maker, you have made that decision to deal internationally.

In the case of extrusion blow molds, some minor tweaking may be required on the molds after initial testing. Do not expect the customer to ship at the expense of cost and timing back over the ocean to deepen a pinch relief area—it's not their responsibility. In several key geographical locations, you have to set up handshake agreements with local tool shops to offer this service.

Quality, value, and delivery are key factors in determining the mold maker of choice, but the one element that holds them all together is service. Be there when needed, 24 hours a day, seven days a week. You should expect your sales engineers to be prepared to make or receive calls in the global marketplace any time of day or night. It will be your decision and, ultimately the customer's, if you prevail in it. You are all doing this for one reason or the other, but ultimately you are gauged by customer satisfaction. If you made the commitment to be the number one best value supplier, then the customers need to be the focus of everything you do. It has to be an obsession and a passion shared by all your employees. Success is based on strong values, a solid commitment to excellence, and strict adherence to integrity.

5.2 TECHNICAL DATA PREPARATION FOR ORDERING A MOLD

The organizational flow or sequence of events in ordering a mold varies from customer to customer. How the mold maker handles this information makes or breaks any project. We should take this opportunity to follow the path of an order from the point at which the mold maker gets involved.

The technical data required to start an order should be attained, by in large, at the quoting stage. Once again you may find yourselves in a situation where personal preferences come into play from customer to customer. Customer A running the same bottle on the same machine could have mold materials or preferred cooling characteristics different, from customer B. It is the responsibility of a seasoned mold maker to speak from experience and certainly not lead the customer "down the garden path."

A qualified mold maker will have prepared a mold specification quote sheet in advance that covers all aspects of the extrusion blow mold itself. Stated earlier, it is the mold maker's responsibility to know the machine and have at its disposal open communication with the machinery supplier. You must be prepared to take this one step further. The mold maker should be prepared to visit with the customer for a technical start-up meeting on all key projects. Although, from experience, this is not always required, nonetheless it is a practice that has proven very successful in the past.

Following are highlights that are the most important features for all extrusion blow molds:

General information
Machine mounting
Mold materials
Mold features
Pinch-off details
Engraving information
Mold cooling information
Mold cavity surface finish
Cycle
Material (product)

5.2.1 General Information

The pertinent machine information is highlighted in this section. This relates to the machine and machine model number on which the mold will run, the number of cavities required, the shut height required, and the center to center distance. This information also includes an article drawing, whether an IGES file is available, and the shrinkage information in three directions.

5.2.2 Machine Mounting

This section covers the various and assorted techniques for mounting the mold to the machine platen.

5.2.3 Mold Materials

This section covers the recommended or mold materials requested for the various components of the mold.

5.2.4 Mold Features

We attempt to highlight the assorted mold features available or required for all extrusion blow molds. Although not limited to these features, we feel this best covers the range from needle blow, to IML, through to striker plates.

5.2.5 Pinch-Off Details

Without a doubt, this is one of the most sensitive features of an extrusion blow mold. In this section we cover the materials used for insertable pinch-offs, whether they are cooled or not, and the style of pinch.

5.2.6 Engraving

This section is very much customer-driven and responds to the client's requirements as far as company logo, recycling logo, the insertable material used, and whether or not it is water cooled.

5.2.7 Mold Cooling Information

This is one of the features, if not the most important, in a blow mold of any type. You must consider the number of systems, the sizes of the water lines, including inlets and outlets, and of course the location.

5.2.8 Mold Surface Finish

This section is governed very much by the plastic material being processed. The sandblast grit, if it is glass beaded (vapor honed), high polished, or if the part requires a texture etched finish, has to be determined.

Using this technical data sheet to your full advantage, you should procure on your own and/or through your customer as much information as possible to offer a firm and accurate quote for their perusal. Depending on the complexity of the quote, you should endeavor to return the requested quotation within 24 hours, preferably the same day. The mold maker should continue to strive to simplify the process by removing layers, even in the quoting process. By responding quickly, it saves your customer time and money in the very critical start-up period.

Suffice it to say that at this point the project has been quoted and the order placed with the selected mold maker. The project begins with furnishing the approved product drawing. In the majority of the cases, you should receive this drawing from the customer, but in many situations you are expected to work with your customer in providing this feature. A qualified mold maker should be willing and able to offer this service to the industry. From the article drawing in collaboration with the blow molder, shrinkage factors are determined and added to the product drawing. Many factors have to be considered in determining shrinkage, not the least of which is the resin being used, the target cycle time, the blow molding machine, the climate of the plant, just to name a few. At this point mold design commences.

At this point the mold maker or, to be specific, the mold designer should have everything required to complete the mold engineering drawings. The mold maker can never be too timid in asking the customer or the machinery supplier technical questions about the project. It is not an admission of guilt. On the contrary, it is evidence that you want to leave no stone unturned. Make your questions good questions but feel comfortable in knowing that the customer would rather have it queried up front than delayed at the end. So as to maximize the production time, your toolmaking department should work in parallel with your engineering department. Experience prevails, and you will know what information should be released to the production floor to get a jump start on the project.

Upon completing the mold engineering, copies are sent to the customer for approval and signing off. Once again, the mold maker should be prepared to accept responsibility for the project because the customer cannot be expected to inspect every detail of the mold design. The customer has its own objectives in reviewing mold drawings. It is the mold maker's overall responsibility for the design and then for manufacture to be blue print specific.

Customarily, in the extrusion blow molding process, it is still common practice to manufacture a unit mold ahead of the production mold set. The rationale behind this is to allow for various and assorted tests before giving the go-ahead on the mold set. The customer could be looking at how the product shrinks, how the part pinches off, the aesthetics of the container, and any stress areas or overall blowing problems. While the unit is being tested, the mold maker can proceed with the production molds and stay metal safe (leaving enough material to incorporate possible cavity, product design, changes before final cavity manufacture) in the areas of concern mentioned previously. Once again, this practice allows minimum interruption in the complete mold cycle and keeps costs under control by running the entire order as a production run. Upon approval of the unit mold, the mold maker can continue to complete the entire mold order. Any mold maker will tell you the difficulty they experience in keeping unit mold costs under control. There is a lot to say about volume orders. Table 5.1 shows the typical technical data required when ordering an extrusion blow mold.

5.3 DESIGN CONSIDERATIONS

A typical blow mold design depends considerably on the product design and of course the blow molding machine. Design of a product to be processed must meet basic requirements, such as capacity, aesthetic or presentation quality, strength, simplicity to manufacture, and easy down-the-line handling to packaging. Complicated design could make tooling cost higher and, on the other hand, the product has to be attractive and presentable to customers.

A typical extrusion blow mold consists of basic components, and depending on the product design and processing requirements, some special components, such as cavity in-

Table 5.1

Technical Data Required for a Quotation for an Extrusion Blow Mold

Details	Pinch-off
Machine	Insert material
Material of part	Cooled
Shut height	Std. pinch-off no.
Center distance	Corrugated
No. of cavities	Corrugated depth
Approved article drawing	Engraving
3-D model	Insert material
Product model	Cooled
Body shrinkage (in three directions)	Company logo
Neck shrinkage	Recycle logo
Type of Mounting	Artwork
Clamp slots	Cavity no.
Overhanging backplate	Mold no.
Runner bars	Asset no.
Material	Cooling line
Backplate	Size
Body	No. of systems
Neck	No. of inlets and outlets
Bottom	Location
Mold Features	Manifold
Blow dome	Cavity surface
Needle blow	Sandblast grit no.
Knockout	Glass bead
Strikers	High polish
Knife plates	Texture no.
Preblow plates	Other
IML	
Detabber	
Weight	

serts, pinch-off plates, interchangeable inserts for different product configurations, have to be integrated in the design.

A typical extrusion blow mold consists of the following elements:

Backplates to hold the mold cavity in place.

Cavity that consists of neck, body, and base.

Neck insert that has the thread finish.

Mold body that is the main part of the cavity.

Bottom insert/base push-up that is the bottom part of the bottle.

Cooling system that regulates mold temperature.

Blow pin/needle that blows compressed air/gas into the parison.

Guide pin and bushing to guide and lock mold halves when closing.

Flash pocket to accommodate excessive plastic material.

Pinch-off to strengthen weld line and trim flash from the part.

Vents to remove air between cavity and parisons.

Engraving insert(s).

Date insert that indicates the date manufactured.

Figure 5.1 is an assembly drawing of a typical extrusion blow mold. Components shown are mold body, bottom, neck insert, pinch-off insert plate, handle eye insert, guide pins and bushings, blow needle, and knock-out pins. Figure 5.2 shows a three-dimensional CAD model of a bottle and parting line generation.

5.3.1 Mold Body

The mold body is the main component of the mold. The majority of the cavity shape is contained in the mold body, including the majority of the cooling system. There can be more than one cooling loop in the mold. In some cases, the cooling system in the mold body is a separate cooling loop. This mold design has separate cooling loops for mold body and pinch-off inserts and the bottom section of the mold. Cooling pinch-off insert plates is recommended if their thermal conductivity is considerably different from the mold body material, for instance, steel inserts and aluminum alloy mold body. This also increases the mold manufacturing cost. Generally, beryllium-copper alloy or Ampco alloy are used for pinch-off inserts and alu-

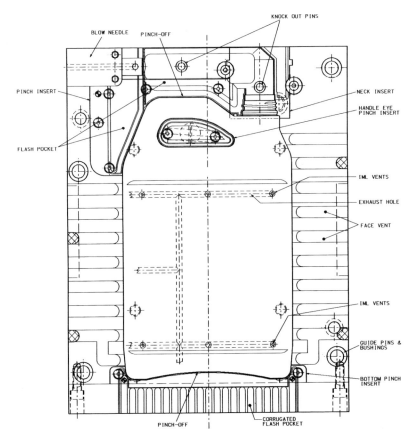

Figure 5.1 Typical extrusion blow mold assembly.

minum alloy for the mold body. These materials have relatively similar thermal conductivities that give homogeneous cooling over the whole mold.

Figure 5.3 shows the mold body and its elements. It has
mountings for bottom inserts, pinch-off plates, ventings, neck
finish insert, blow needle/pin, and handle eye inserts. Normally guide pins and bushings are located on the mold body.
When designing a mold, all of these components and function-
al elements have to be integrated in an optimized order. Basi-

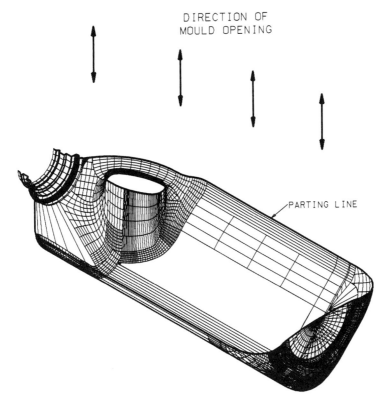

Figure 5.2 3-D CAD model of a bottle and parting line generation.

cally, the mold body is split into two halves, pin half and bushing half, (upper half and lower half) at the parting line.

Most container molds have flat parting lines because they are normally circular in shape or are symmetrical. In some cases, the parting line is not necessarily on a single plane. Generally speaking, the parting line is the section generated in the direction normal to the mold opening, that means, in the direction where there is no undercut. As a simple definition, undercut is defined as an area or areas where one cannot see within the parting line profile of the cavity if

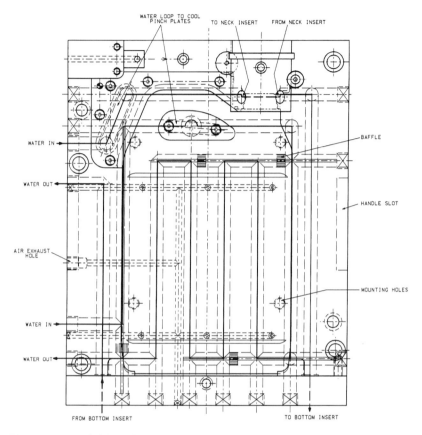

Figure 5.3 Mold body and its elements.

looking in the direction normal to the parting line plane. In
some cases depending on the shape of the article, there can be
undercuts.

Care should be taken at the parting line area on the mold
because it can be chipped very easily. Normally wear-resist-
ant material, such as steel or beryllium-copper, is used at the
parting line area for better durability, for instance, at pinch-
off areas, such as a handle, shoulder, and bottom flash areas.

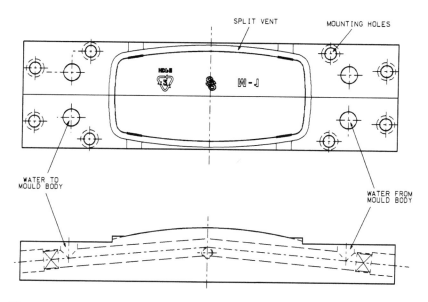

Figure 5.4   Mold bottom section and elements.

5.3.2   Bottom

Most container extrusion molds have a bottom (base) section for mold manufacturing or processing purposes. Figure 5.4 shows the bottom section of the mold and its elements. It has a separate cooling loop for the bottom together with the pinch-off inserts from the mold body. Bottle identification engraving, customer logo, date, and mold cavity numbers are engraved on the bottom cavity surface.

5.3.3   Neck Finish

In the extrusion blow molding process, some of the neck finishes are molded during the blow molding process, and some are preformed. Threading is done on a thread insert. Figure 5.5 shows a neck insert which forms threading for closure of the container. It is also important to have a good pinch-off on the threading area. Pinch-offs at the threading insert fail

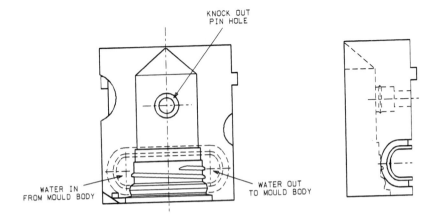

Figure 5.5 Neck insert that forms threading for closure of the container.

sooner than other areas. Therefore, threading is done by using an insert, and this is easily replaced when required. This also allows interchangeable inserts for different finish designs. Depending on the type of blow molding machine blow nozzle or blow pins, their locations are different. It is also important to cool these inserts properly because this area normally is thicker and hotter than other areas.

5.3.4 Cavity Inserts

Cavity inserts in the extrusion blow mold, are not as common as you may find in injection stretch blow molds. They are employed for different styles of the bottle, such as instructional engraving for the product packed in the bottle, to produce different products from the same mold with slight variation, because of a complex cavity shape which could not be manufactured in one solid block, because of considerable amount of undercut which needs to have a sliding cavity insert, or because of a peculiar product design which needs a sliding cavity insert for better processing.

5.4 EXTRUSION BLOW MOLD MATERIAL

The quality of blow molded parts depend primarily on the mold and blow molding machine. Durability of the mold is also important depending on the number of parts to be produced from that particular mold and the number of parts per hour which the mold can produce. All of these factors depend largely on the design and mold material. Therefore, the mold design and the material selection are based on the overall sales volume of the molded part. It can also be evaluated by a cost analysis and also on the basis of particular plant practice.

The following materials are most common in making extrusion blow molds (see Table 5.2).

5.4.1 Aluminum Alloy

A number of different aluminum alloys are used in producing extrusion blow molds. The following materials are preferred blow mold materials by experienced mold makers.

Table 5.2

Typical Extrusion Blow Mold Materials and Properties

Material	Thermal conductivity (Btu in./ft^2 °F)	Density (lb/in.^{3})	Hardness	Tensile Strength (psi)
Be-Cu alloy				
CA 172	770	0.298	37 Rc	150,000
CA 824	750	0.304	34 Rc	135,000
Al alloy 7075-T6	900	0.101	14 Rc	73,000
Al alloy 6061-T6	1140	0.098	8 Rc	68,000
QC-7	1090	0.102	17 Rc	80,000
Ampco 940	1500	0.315	20 Rc	100,000
Ampco 945	960	0.310	31 Rc	130,000
AISI P-20	200	0.282	30–36 Rc	120,000
AISI 420 stainless steel	166	0.280	50 Rc	213,000

Aluminum alloy 7075-T6 is a general purpose high-strength aircraft grade alloy which contains zinc, magnesium, copper, and chromium. It has very good thermal conductivity, excellent machinability, light weight, and corrosion resistance.

QC-7 is a high-strength aluminum alloy that is fully heat treated and stress relieved. It has very good thermal conductivity, high strength, and surface hardness. Therefore it is suitable for polishing and texturing. It can also be welded and has very good machinability.

5.4.2 Beryllium-Copper Alloy

Beryllium-copper alloys are used extensively in blow mold construction. It is tougher than aluminum alloys and also has excellent thermal conductivity and corrosion resistance. Beryllium-copper alloys are about three times heavier than aluminum alloys. Therefore it is more expensive than an aluminum alloy. Its drawbacks are poor machinability and high cost.

In most cases beryllium-copper alloy is used where higher toughness is required, For example, in the pinch-off area in the extrusion blow mold. Normally, it is used as an insert, where hardness and corrosion resistance are required. It is also important to note that beryllium-copper alloy dust produced while machining may cause lung damage.

Ampco(R) 940 and 945, alternatives for beryllium-copper alloy, are nickel-silicon-chromium-copper alloys developed by Ampco. They have very good thermal conductivity and favorable hardness. In addition to these properties, they accept etching and texturing, are weldable with other copper alloys, and have good corrosion resistance, and average machinability.

5.4.3 Steel

Steel is widely used in making PET, PVC, or engineering resins bottle molds. It is extremely tough and corrosion resist-

ant. For decorating a cavity surface, it provides excellent texturing capability. Its drawback is poor thermal conductivity.
Steels used in this application are AISI P-20 and stainless
steel. In some cases, processors prefer to use H13. Machining
these materials is difficult because of their hardness, but they
gives longer mold life. However, because of extreme competition in introducing new products in the market, the product
life cycle of the containers produced by the mold is shorter
than the life of the mold itself. Normally, combinations of aluminum alloy and steel are compromise solutions in this type
of mold.

5.5 COOLING

Cooling time can take up to two-thirds of the mold closing
time. Therefore, it affects the overall molding cycle time. To
produce higher number of parts per hour mold cooling has to
be effective. Cooling is done by cooling the mold block and by
internal cooling. Cooling is based on the coolant flow rate,
coolant temperature, and overall heat transfer efficiency from
the hot parison to the coolant flow.

Mold makers are often requested by customers to design
cooling systems in a particular manner, but in most situations
they design because their years of experience lead in this
field. Nevertheless, the customers' or the end users' input is
vital.

You have to consider, however, whether these molds run
on a modern piece of equipment or on one that is basically
outdated by today's standards. In other words, you have to ask
yourselves if a mold that is cooled optimally is necessary for a
30-year-old molding press. Cooling should be optimum, but
suited to the particular machine because a cost factor is involved for any upgrading.

For one, not all plants have climate-controlled production
facilities and an overdose of cooling can makes the mold sweat
under these conditions. Consequently, the cooling system is

choked down to a trickle so that the extra cost for a high production mold was spent in vain. Low labor costs and the pace of production is another factor in Third World markets and can be considered when developing a cooling system for a mold that runs under these conditions.

In some cases mold makers are asked to provide maximum cooling which is relative in comparison to that mentioned previously. To be fair, you need more information regarding the purpose of this request. A customer who wants cooling lines every inch around the cavity can, in most instances, be convinced to go by a more common route in accord with our normal design procedures.

The stigma, the more the better, originates from the European molds that were made of steel long after North American molds were produced from forged aircraft aluminum alloy. To explain this further, it is of no use to pack the molds with cooling channels in areas where the material of the blown container is the thinnest when it is impossible to cool places, such as flash areas, adequately where cooling is needed the most. It is easy to understand that the thinnest part always must wait for the heavier sections to cool off to avoid distortions of the molded article in certain areas.

Container molds are usually built in three sections: the top section with the neck finish insert, the body, and the bottom section. Every section has its own individual cooling circuit that allows control of the cooling medium within this area, more cooling in the flash areas and less in the center section where there is no heavy accumulation of plastic material.

This is even more important when we realize that it has become common practice to insert the pinch-offs on aluminum molds with beryllium copper or steel for longer lasting pinch-offs, depending on the type of plastic involved in the molding process. A decision has to be made whether to run a cooling line directly through the insert, which is usually only possible on larger molds, and have a milled channel butted up against the insert sealed with an O-ring (see Fig. 5.6), or as it is done

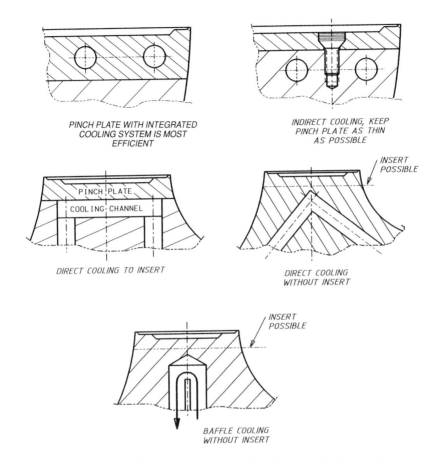

PINCH PLATE WITH INTEGRATED
COOLING SYSTEM IS MOST
EFFICIENT

INDIRECT COOLING, KEEP
PINCH PLATE AS THIN
AS POSSIBLE

PINCH PLATE
COOLING CHANNEL

DIRECT COOLING TO INSERT

INSERT
POSSIBLE

DIRECT COOLING
WITHOUT INSERT

INSERT
POSSIBLE

BAFFLE COOLING
WITHOUT INSERT

Figure 5.6   Cooling line through insert that is sealed with an O-ring.

in most cases with indirect cooling where a cooling line runs
beneath the insert. Here the inserts should be made as thin as
possible for the least loss in heat transfer. It is also important
to have a good surface finish between the two metals because
any air gap is an insulator.

Now, it is crucial that the experienced mold designer
makes his layout in so that the maximum cycle time for a particular container is achieved. Computer software is available

to backup the designer in his judgment of the right cooling channel diameter, distance from the cavity, and spacing from each other, based on the influx of heat from the cavity surface.

Another trend is the large diameter cooling line, but we are seeing less of milled cooling channels sealed with a plate and an O-ring. The latter is still used on large molds made of cast aluminum or steel that have large chambers cast into the back of the mold. This type of mold is not very well suited to drilled lines due to the porosity of the materials. This has often been circumvented by placing a copper pipe around the cavity which then becomes the cooling channel once cast. The cooling qualities are poor even here and the cast mold is on the way out because these molds cannot keep up with the speed of modern blow molding equipment.

It is easier to place cooling channels into a larger than into a smaller mold, especially multicavity molds. The concern is always the same. The machine platens of the molding machines are too small for the mold that has a large cavity. What suffers is the distribution of cooling lines in the mold block. There is nothing much that can be done here. One cooling line might have to pick up heat from as far as two to three inches. Although this is not desirable on a small cavity mold, it might be the right approach to a large container mold. We have seen molds built in Europe as recent as 1997 that have a close formation system where all cooling lines are one inch apart from each other, running as close as 3/4 in. parallel to the cavity diameter, and they are connected with milled channels from both sides of the mold block. This is not only more expensive to produce but also cause for concern about possible leakage.

This is overkill in an aluminum mold and, as mentioned earlier, was taken over from steel mold designs whose heat transfer is three times less than in mold grade aluminum.

No doubt, this type of cooling is even around the cavity, but this is also the case when a system is fitted circumferentially around the cavity diameter with angle lines.

It becomes clear when we look at the "cooling cone" or the angle of radiation from the cooling line (see Fig. 5.7). Even if the cooling line diverts from the cavity, as a straight line does

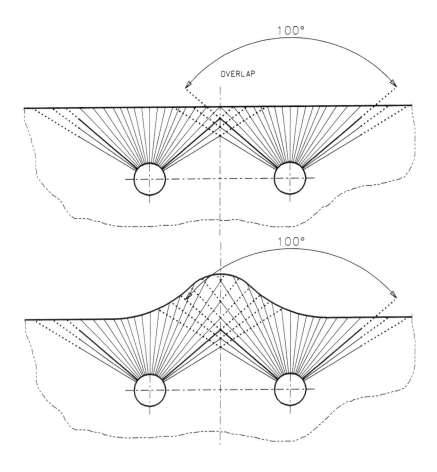

THERE IS NO NEED FOR CONCERN IN ELEVATED AREAS WITHIN
A CAVITY AS CAN BE SEEN HERE.

Figure 5.7   Angle of radiation from the cooling line ("cooling cone").

from a round cavity, now cooling is supplied from both sides, and at the points where the lines join, turbulence is generated to make the effect even more substantial.

One might even be concerned that the cooling water temperature on long systems might have an effect on the "in" and "out" temperatures and so influence shrinkage of the molded part. Two systems might be recommended where each system covers only one-half of the mold or a system within a system where every second line belongs to the second system (see Fig. 5.8).

The trickiest and most educating question is surely how a cooling channel should be sized and how close it should be placed to the cavity for maximum efficiency in a high-performance mold. Cavities are not always even so that a great deal of judgment is left to the experienced designer. But some guidelines can be recommended. Here we want to have a closer look at the "cooling cone" or radiation angle mentioned earlier. It is best explained by a layout showing how it should be done and what should be avoided.

Cooling channels too close to the cavity and spaced too far apart can have an impact on the shrinkage of a blown article in reference to cold or hot spots. The "cooling cone" graphs show the relationship between these two.

Even though cooling radiation is distributed circumferentially, we have to observe the side that is directed toward the cavity. If fewer lines are used for a cut rate or low production mold, it would be better to place them further apart from the cavity so that the "cooling cone" becomes smaller and al-

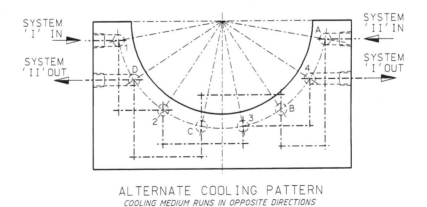

ALTERNATE COOLING PATTERN
COOLING MEDIUM RUNS IN OPPOSITE DIRECTIONS

Figure 5.8 System within a system.

lows the mold material to be cooled more evenly rather than spottily.

If the "cooling cone" is a true measure of good cooling and we apply this to a round cavity, then we will be surprised at what we discover. Now, we can place the drilled channels further apart according to the overlap and still get the same cooling performance. The emphasis is on overlap.

It is important to remember that all modern blow molding equipment runs faster and more efficiently with parison control, and the mold maker has to follow suit with mold materials and cooling distribution.

5.6 VENTING

Extrusion blow molding process is a process that forms a heated thermoplastic parison to a female mold cavity shape by blowing air or gas into the parison after the mold closes. In the process air trapped inside the mold has to be released from the mold cavity so that the parison conforms to the female cavity shape effectively. Poor venting in a mold can cause various problems in blow molding, such as sink marks, a rough molded surface, rocker bottoms, thin corners, indented parting line, and incomplete blow. All of these result in undesirable molded articles. Therefore, good venting is also a very important part of the mold design. All of the air between the cavity and the parison must be removed completely as the part is blown. If any air is trapped in between the product and the mold, the part will have a pitted surface, it can be rough in appearance, and the air causes poor cooling of the part. The total parison must be completely in contact with the cavity surface. Vents are normally located in the following areas of the mold:

Parting line
Complex shape corners
Deepest area of the cavity
Periphery of the cavity insert(s)

Corner radii
Bottom base split
Neck area

Basically, venting should be provided where air could most likely be trapped. If possible, the cavity should be sandblasted for improved venting. There are several ways to vent a mold cavity. The following methods are very common in typical extrusion mold design.

5.6.1 Design Integrated Venting

These venting systems are the most favorable method. Typically, vents are at the parting line area, called face vents, between any splits of the cavity parts, and around any cavity inserts. These are primary venting locations, and there are collectors designed for these vents to exhaust from the cavity during each molding cycle. Figures 5.3 and 5.4 depict these venting systems.

5.6.2 Vent Plugs

These are used in the cavity surface where venting is required. They are located mostly in the deepest area of the mold halves, in sharp corners, in specially decorated areas, and in complex cavity shape areas where the air is most likely to be trapped. Vent plug size varies depending on the venting requirement and the blow molded part surface finish requirements. Figure 5.9 shows typical vent plugs used in extrusion blow molds.

5.6.3 Pin/Spot Vents and Split Vents

These types of venting systems are employed occasionally in an extrusion blow mold. The size of a pin/spot vent varies from 0.020 to 0.060 in. and for split vent widths vary from 0.007 to 0.020 in. and lengths from 0.1 to 1.0 in. depending on the manufacturing process. These vents are machined by

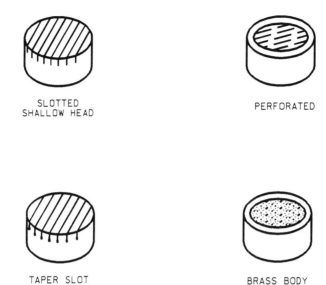

SLOTTED
SHALLOW HEAD

PERFORATED

TAPER SLOT

BRASS BODY

Figure 5.9   Typical vent plugs used in extrusion blow molds.

drilling or electrodischarge machining methods from the cavity side to the collectors in different parts of the mold. These
venting systems are particularly useful in small corner radii
and in complex cavity shapes.

5.7   CAVITY

The mold cavity is the area where the heated parison is
formed into its desired shape by the blow molding process.
Therefore, it is one of the most important elements in mold
design. In general, the article or product design is done by the
processor or packaging company based on market demand,
packaging requirements, presentation, and physical and
chemical requirements. There are several ways to design the
article to be blow molded and deliver it to a mold maker to
manufacture a mold.

5.7.1 Cavity Design

Physical Model

Most processors are still making physical models of articles to
be blow molded by the wood pattern making process. The advantage here is that you see, feel, and decide whether the
model made by the pattern maker represents the article you
want to produce via the blow molding. The actual volume of
the model can be determined by displacement. The displacement method measures volume by placing the model in water
and measuring the amount of water the model displaces. The
wooden models can be finished within 0.002 in. and wave used
to manufacture a mold cavities before the high-end CAD/CAM
software packages were introduced. Although an outdated
process, it is still used in many instances.

3-D Surface/Solid Model

As technology advances, more and more processors design
their products using computer-aided design and manufacturing software (CAD/CAM). Products are designed in either surface model or solid model in CAD/CAM software, and cavity
information can be transferred electronically to a mold maker.
There are a number of advantages in this process of making a
cavity design. Product design can be analyzed in all aspects
within the design stage. Volume can be measured accurately.
The model can be generated into a finite-element mesh and
the blow molding process can be simulated and analyzed. The
final cavity surface/solid model can be used directly to generate the CNC programs to manufacture the mold cavity. The
mold designer can utilize cavity surface/solid model to design
the mold directly in his CAD software.

5.7.2 Cavity Surface Finish

Mold cavity finish requirements depend on the thermoplastic
to be processed and the type of finished articles required. It is
important to have the proper finish on the cavity surface because the cavity surface is duplicated on the article to be

molded. There are four main categories in extrusion blow mold cavity finishes, sandblasted, glass-beaded, polished, and custom-etched.

Sandblasting (HDPE)

The cavity finish must be completely benched before, sand-blasting. Of the four types of cavity finishes, this is the most widely used and, in some cases, is combined with the other three. Sandblasting is done with various sizes of grit. Ranging from small sand particles to small rocks, according to the cavity size. Containers from 2 oz to 55 gallon drums and larger need to be sandblasted because the cavity surface must have a rough surface to help in removing the air when blowing the container. The larger the container (in size or capacity), the larger the sand or grit. Grit has a standard numbering system. The smaller size sand has a higher number and the larger sand particles have a lower number (grit #80 is a fine sand and grit #16 is a rougher, coarse sand). Venting of the cavity is very important and covers minor impurities like mold sweating in higher humidity conditions that give the outside of the container a matte finish.

Glass Beading (HDPE, PC, PET)

The Cavity finish has to be semipolished and smooth before glass-beading. This type of cavity finish is done mainly for venting, not for changing the outside surface of the container. The container will have a glossy surface look and maintain a clear appearance. Glass beads come in one form as are small beads of glass that explode on contact with a hard surface and create a frosted appearance on the cavity surface.

Polished Cavities (PET)

This finish requires that the cavity to be highly benched and polished to create a glass-like appearance on the surface of the container. If required, light glass-beading may be used to help with venting of the cavities. There are various levels of polish, but most processors prefer a mirror-like surface.

Texturing

Texturing of a cavity is usually used to a specific look and feel to the product line. It ranges from a wood grain to a rough surface with small indented areas. After the texturing has been completed, the product is usually sandblasted with a fine grit to clean it up.

5.8   PINCH-OFF DESIGN

The Pinch-off section of the mold is where the heated parison is squeezed and welded together. Pinch-offs carry out three major functions in blow molding. The pinch-off has to be tough enough to withstand pressure of mold closing and opening cycles. It has to be designed so that it pushes small amounts of material into the cavity for a stronger weld and at the same time cuts off the flash from the main body of the blow molded article. To accomplish this, most molds have a double pinch or a compression area. The flash pocket depth is related to the pinch-off area and is very important for proper molding and automatic trimming of the part. Flash pocket depth is determined by plastic density, weight of the part and the parison, and the parison diameter.

Inappropriate flash pocket depth leads to many undesirable results. For instance, a shallow pocket can cause extreme pressure at the parting line and machine press section and can also cause difficulties in trimming flash from the part. A deep pocket can cause poor cooling in the mold and a hot flash can soften the weld area and weaken it.

It is also important to cool plastic as fast as possible to facilitate the subsequent operations after the molded part is ejected from the mold. Therefore, good thermally conducting material, such as beryllium copper or Ampco(R) Copper must be used. Bottle blowout can occur during molding if the temperature around the pinch-off area is too high.

The pinch-off design varies depending on material to be processed, the type of blow molded part, the size of the molded article, and the molding process. The following are typical pinch-off designs:

5.8.1 Single Pinch-Off

Most small container molds use a single pinch-off design or single pinch-off with a compression area to strengthen the welded area. There are slight variations in this type of pinch-off design depending on the thermoplastic to be processed, the blow molding machine, the bottle design, and the molding process. Figure 5.10 shows single pinch-off design variations. The pinch-off landing for smaller bottle molds (up to 5 gallons) varies from 0.010 to 0.040 in. Normally, the pinch-off landing width on one-half of the mold differs from the other half for better cutoff action.

5.8.2 Double Pinch-Off

A double pinch-off is normally used in medium to large industrial blow molds. There are different types of double pinch-off designs and their variations are based on processing requirements. Some of the double pinch-off designs commonly used are shown in Fig. 5.11. Primary and secondary pinch-offs widths differ from mold to mold. Normally, primary pinch-offs in large industrial extrusion blow molds are at the parting line level and widths vary from 0.040 to 0.080 inch. Secondary pinch-offs are below the parting line by 0.010–0.040 in. and

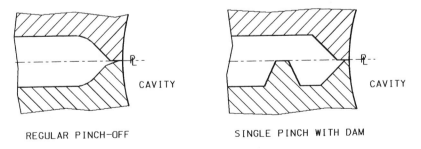

Figure 5.10 Typical single pinch-off variations.

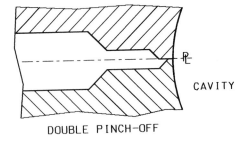

DOUBLE PINCH-OFF

Figure 5.11 Typical double pinch-off variations.

the widths are equal to or smaller than the primary pinch-offs.

5.8.3 Special Design

Figure 5.12 shows variation of custom pinch-off designs. The design is normally tailored to particular processor and processing requirements.

5.8.4 Flash Pocket

The flash pocket primarily accommodates excess thermoplastic parison after the mold is closed. At the same time, the flash pocket cools the hot thermoplastic and transfers the heat through the cooling system. It has to be just large enough to accommodate excess thermoplastic. A shallow flash pocket leads to improper mold closing and causes other processing

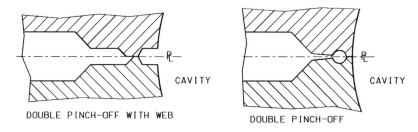

DOUBLE PINCH-OFF WITH WEB   DOUBLE PINCH-OFF

Figure 5.12 Customized pinch-off design variations.

problems, such as improper pinch-off weld, excessive pressure on the machine platen, and trimming problems on trim tooling stations because the molded part is not properly pinched. Too deep a flash pocket results in poor cooling of the flash, weak pinch-off weld, and trimming problems.

Therefore, the flash pocket has to be designed properly to avoid these problems. The depth of the flash pocket and volume required can be calculated on the basis of parison diameter, parison thickness, preblow pressure, and parison orientation. The heat transfer efficiency from the hot parison to the mold cooling system is directly proportional to the flash pocket surface area. To achieve a larger surface area, a corrugated flash pocket design is recommended. Figure 13 shows different variations of the flash pocket.

Improper pinch-off and flash pocket design causes processing problems, which are listed in Table 5.3.

5.9 MANUFACTURING TECHNIQUES

In this section, two main areas of manufacturing techniques are discussed in mold cavity manufacturing and complete mechanical work on the mold. The following techniques are employed in a typical mold manufacturing process.

Stress relieving is a process to reduce internal residual stresses in a metal object by heating the object to a suitable

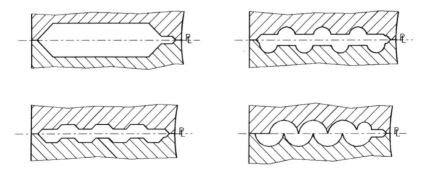

Figure 5.13 Typical cross-sections of flash pockets.

Table 5.3

Improper Pinch-Off Design vs. Processing Problems

Problem	Cause
Flash tearing off	Long pinch-off landing
Part blown out	Pinch-off too hot
	Pinch-off too sharp
	Pinch-off too wide
	Pinch-off gap too wide
	Flash pocket too deep
Weak weld line	Pinch-off too sharp
	Secondary pinch-off too deep
	Flash pocket too deep
	Poor venting
	Pinch-off gap too wide
Improper parting line weld	Flash pocket too shallow
	Mold misalignment
	Inconsistent pinch-off gap

temperature and holding it for a proper time at that temperature. This treatment is used to relieve stresses induced by casting, quenching, stamping, machining, or welding.

Turning is a process that removes metal by using a stationary tool on a rotating workpiece. Most of cylindrical shapes required in a mold are produced by turning on an engine lathe or CNC (computer numerically controlled) lathe. It is a fast metal removal method, and normally the surface finish is better.

Milling is a process that removes metal by rotating cutting edge or edges on a stationary or moving workpiece. The Cutting tool is guided in a coordinated movement while the workpiece is fed past the tool. Milling is a universal complex machining method that incorporates advanced technology, such as computer numerical control, servo systems, cutting tool technology, and complex machinery. Today's fast delivering mold manufacturing capability would not be possible without advanced milling operations.

Drilling is a hole-making process done by rotating tools on a workpiece. Relative movement of the two feeds the tool into the workpiece. In general, all cooling lines are manufactured by drilling operations. Which must be done before tapping and premilling.

All threaded holes are made by *tapping* predrilled holes. Plugs, mountings, and fasteners for mating parts in mold manufacturing are produced by tapping.

Boring is used primarily when close dimensional tolerances and good finish are required. Normally boring is done on predrilled holes. It is used to make guide pins, bushings holes, and sliding insert guides in a mold.

Grinding is a process of machining by using abrasive grains to shape a workpiece. This includes cleaning of the surface, shaping, sizing, surface finish improvement, and separating, such as cutting-off or slicing. Grinding methods include cylindrical grinding, surface grinding, internal grinding, and tool grinding.

Electrodischarge machining (EDM) is a machining process that produces cavitation proportional to the electrode shape in an electrolyte on a work material by electric discharges between the two. Cavity shapes Difficult to cut in machining centers and highly precise small cavities are often manufactured by the EDM process.

The Manufacturing process and cavity manufacturing techniques also depend greatly on product design. A simple cavity can be machined by simple turning or milling whereas a complex cavity has to be partly produced in a machining center and partly by EDM or even on five axes simultaneously.

5.9.1 Cavity Manufacture

Because most cavity shapes differ for different applications, there are many different cavity manufacturing techniques. For instance, cylindrically shaped cavities are best manufactured by turning on lathes whereas complex cavities are produced in advanced machining centers. On the one hand, highly precise small cavities are ideal for the EDM process. On the other hand, today's advanced high-speed machining centers can do even the EDM work. Decision in the manufacturing method chosen is the most economical, fastest and most efficient method that meets delivery requirements.

In software, CAD/CAM application in mold making has been growing rapidly as technology advances. Today, blow mold makers deal with customers who want their orders delivered in two to six weeks depending on the type of mold and the number of molds to be manufactured. With this kind of constraint, mold makers cannot rely on conventional mold making processes. Mold design, cavity design, cavity machining, and mold block manufacturing have to be done almost simultaneously to meet delivery or the customer will look for a new mold maker. Therefore CAD/CAM application is inevitable to facilitate mold making. Cavity design has to be done first and then mold design, cavity manufacturing, and all mechanical work follows. At this point sharing the same data plays a vital role in achieving accuracy, effective time management, and manufacturing.

The design procedure is to analyze the shape of the cavity. Some important check points are the depth of the cavity, whether standard cutting tools available can be used in manufacturing, whether the cavity shape at the corner radius is large enough to be machined by ball nose or special tools, whether there is an undercut and whether a special cutter can be used or different setup is required (angle setup), or whether to use additional axis in the machining center (i.e. fourth axis or even fifth axis). If the cavity shape cannot be machined, then use EDM to burn the cavity. Most bottles, especially injection/stretch blow molded parts, are round and can be turned on a lathe. If the mold is turned on a lathe the surface finish is much better than from a machining center and is also produced faster.

Reverse Engineering

Even though the application of CAD/CAM technology in mold making has been growing rapidly for the past decade, there are still many needs for reverse engineering. Following are some of the reasons:

Customers choice of pattern making and models for the product they required

Artwork required for a certain design that is easier to make by hand

To duplicate the exact shape of an existing product

Basically two popular methods are employed to convert a physical model to electronic data, noncontact and physical contact digitizing. The former is done by laser scanning and the latter by mechanical scanning.

Mechanical scanning uses a mechanical probe to digitize the model. The Scanning direction selected depends on the geometry of the model. Preferably you scan in the same direction as if you produce a tool path for the particular shape to get the best representation of point cloud data which best represent the model. Roughly speaking, you may want to scan so that the scanning direction is perpendicular to the shape being digitized in the same way as you would program your tool path for the best surface finish. It may not always be the case, however. Therefore, you may have to digitize several localized areas in different directions. This gives the best representation of point cloud data for the model. Unfortunately most reverse engineering software packages support single direction scanning data to create a surface model. Therefore it has to be processed individually and then combined.

In laser scanning, point cloud data represent the actual shape scanned. In mechanical scanning, point cloud data represent the trajectory of the probe center or the tip of the probe. This can be processed when you read the point cloud data into the software to get a surface model of the actual shape. It has to be processed before creating the surface.

A number of software packages are available in the market today. The system best suited for your application depends on how complex a shape you have to work with. If the shape is very complex, most reverse engineering software packages may not give you a push button method. There will be some manual manipulation for surface modeling. You also need good surface modeling and styling software.

If there is no need to create a 3-D surface model, there are also systems which digitize and produce tool paths within the system. This may be a good solution if further analysis of the model is not required. e.g., volume analysis, flow, stress, or simulation.

In designing parts for blow molding, some customers prefer to cut a model and feel it. Then they hand finish the model to the desired shape. As a mold maker you must be able to make cavity from any cavity information that your customer supplies. Therefore, reverse engineering is an important part of the mold making process.

5.9.2   Mechanical Operations

To be efficient in manufacturing small capacity bottle molds with large number of extrusion blow mold manufacturing process, it is mandatory to reduce setups as much as possible. The number of setups also depends on the complexity of the design. In most of blow mold manufacturing processes, mechanical operations, such as drilling, tapping, boring, grinding, pocket milling, threading, facing, chamfering, take between 60–75% of the manufacturing time. Therefore as many operations as possible must be done in each setup. Combining different operations at a workstation is desirable, for instance, a turning machine with milling capabilities, multiaxis machining, five-axis milling and boring machine. It is also important to note that mold design has to consider manufacturability.

5.9.3   CAE in Blow Mold Design

Typically extrusion blow mold design and manufacturing is based on experience and trials. In most cases, when developing a new product, there are a number of modifications and revisions on the cavity and mold designs to achieve optimum processing. It is a very expensive product development process. A number of companies are developing software packages to analyze plastic behavior during the blowing process, and a limited number of simulation software packages are available for analyzing design and processing.

The Type of analyses applicable to blow mold making are flow analysis and cooling analysis (thermal analysis). Basically, these analyses fall into computer-aided engineering. Analysis is a tool but not an automation technique. It can be done before producing an expensive design to estimate the outcome of a particular design. This saves precious time and costs involved in mold making.

The basic methodology behind the CAE technique is that a design or process is proposed as the first step. Then, the engineer constructs a model, or representation, of the specific design using the method prescribed. The computer rapidly evaluates the results of both the input conditions and the model which the engineer described. Then, the output conditions are listed by the computer, the engineer evaluates the consistency of the results based on his experience and then determines how the design has to be modified to achieve acceptable results. The process is repeated until a succesful design is achieved.

Research into plastic flow has been popular for a few decades especially in injection molding. However, research in blow molding has not been done as much as required because many of the parameters are difficult to analyze. Recently a few analysis software packages became commercially available. However the area covered is mainly injection/stretch blow molding. Intensive research is being done in industrial blow molding especially in the automotive industry in North America driven primarily by the big three auto makers of North America.

Effective, balanced, and sufficient cooling is very important in blow mold making. Excessive cooling or insufficient cooling may affect the blown part. Optimizing cooling reduces manufacturing cost and improves the product. It also saves cycle time, part costs, and try-out costs. The main benefits are quality and productivity.

Product quality is based on the accuracy with which the mold cavity is fabricated, the repeatability of molding machine used, and the appropriateness of the mold design. Part of correct mold design is a cooling system that extracts heat from the melt in the mold cavity at the maximum rate possible and uniformly throughout the mold. Another major benefit of effective cooling system design is the impact on productivity. A Mold designed with optimum cooling produces parts in the shortest possible time. This produces large cost savings in part production and a less expensive and better product.

5.10 INSPECTION OF A BLOW MOLD

It should be mandatory practice to inspect all components of the extrusion blow mold. Granted, today's technology and the repeatability of CNC machinery eliminates the likelihood for error from mold to mold. One can never underestimate the importance of a thorough inspection. At Wentworth, every component of every mold is inspected and *any* discrepancy and is adjusted immediately.

It may seem overly dramatic to say that we check every dimension. The fact is that the mold maker should know what the critical dimensions are for the customer and for the machine builder. As in any tool making application, tolerances and open dimensions should be considered. A water line that is out 0.010 in. versus a shut height that is out 0.010 in. do not even compare. Any reputable mold maker knows this and reacts accordingly.

What is of vital significance is that the inspection covers the machine mounting and cooling characteristics. When the mold arrives at the customer's plant, it must mount on the machine and run the first time every time. Once again, a good working relationship with the machinery supplier puts the mold maker in a position to have these characteristics on hand and available before mold design. In some cases machines have been retrofitted and altered to a customer's specifications. Thereby, the old rule of a technical meeting before start-up is once again strongly recommended.

Table 5.4 is an example of a detailed inspection report for a mold body of a standard extrusion blow mold. This report covers the critical dimensions of the mold as it relates to the cavity and, therefore, the bottle and also, once again, the mold mounting to the machine.

As seen in the report, the following are key characteristics:

1. Mold shut height: mold width, length, and height
2. Water flow and water line pressure check
3. Mold venting dimensions
4. Neck dimensions
5. Pinch-off depth, land, and width
6. Engraving information
7. Center to center dimensions

Although not limited to these details, they are certainly some of the more critical for the extrusion mold body. Depending the individual customer needs, the mold maker should be prepared to offer this detailed inspection report to the customer at any time. At Wentworth, we make it common knowledge that the report exists and many customers take us up on our offer by requesting that we include the report with the mold shipment. Many customers ask that we customize confidential style reports for their individual needs, and we are only happy to oblige. Many other vital reports should be provided through CMM checks, volumetric checks (both conventionally and through a 3-D file), and cavity data via digitizing technology (see Table 5.4).

5.11 MOLD MAINTENANCE

"So, how long will your blow mold last?" If every mold maker had a nickel each time this was asked of them, we would never have to worry about a delivery date again. Mold mainte-

Table 5.4
Sample Inspection Report

Date:		Inspection Report:		Checked:
MSO:		Customer:		P.O.:
Style:			Works:	
D.W.G.				

	Dwg. Dim.	Mld. #1	Mld. #2	Mld. #3	Mld. #4	Mld. #5	Mld. #6
Graduation							
Cust. Stampings							
MSO and mold No.							
M. 1/2 Height U.							
L.							
Upper cut-out							
Over top							
M. width							
M. length							
Neck bore							
Cavity h. to neck S/L							
Cav. h. neck S/L to bottom S/L							
I.M.L							
Cav. width B							
Cav. depth T.B.							
Cav. depth T.P.							
Cav. depth C.B.							
Cav. depth C.P.							
Cav. depth B.B.							
Cav. depth B.P.							
Bushings and pins							
Venting							
Water flow							
Water leaks							
Sandblast							
Mounting Upper							
Lower							
Keyways Upper							
Lower							
Pinch-off depth top							
Pinch-off land top							
K.O. system							
Angle Top							
Bottom							

nance and care are probably the most underestimated aspects of the mold cycle. We can't answer the question just asked if the blow molder does not have a preventative maintenance program in place and does not handle or care for the molds in the manner they require.

Probably the area of most concern to the mold maker is how the blow molder looks after the molds in between a changeover. The following list accents what we feel enhance the life of the tool if followed properly.

5.11.1   Blow Out the Water Line System

Using compressed air, remove the in and out water fittings and methodically blow out the water line system. Any remnants of water/coolant or corrosion build up must be removed before putting the molds on the shelf for storage. This is of particular importance when storing or transporting the molds in extremely cold climates.

5.11.2   Blow Out the Drilled Venting System

Using the same method as above, take the time to blow out the drilled venting system. Remove several plugs so as to get at the system from several locations to remove any trapped dirt, residue, or moisture.

5.11.3   Clean the IML System

Once again, follow the same routine as above to ensure a clean system. If the blow molder is using a slotted vent plug for the IML (inter mold labeling) vent in the cavity, use the corresponding "feeler gauge" to open up the slots.

5.11.4   Replace O-Rings

If the molds are going to sit on the shelf for an extended period of time or if the molds have been running for a prolonged period, it is advisable to replace the O-rings. Depending on the coolant used by the blow molder, they become worn and experience some fatigue. Use this downtime to replace all of your O-ring seals.

5.11.5 Replace Pin and Bushings

Pin and bushings should not be strictly responsible for guiding the two mold halves on the machine. The machine platens and accuracy of the molds themselves should assume most of that responsibility. The parting-line pin and bushings, though, wear depending on the life of the machine and the molds. Once again, seize this opportunity to change all your pins and bushings during this period.

5.11.6 General Wipe Down of the Mold

After all of the above, "blow off" with compressed air and entirely wipe down the blow mold with a soft cloth.

5.11.7 Spray Oil in the Mold

Last but not least, using a greaseless lubricant (there are many on the market), spray the entire mold, cavity included, to protect the mold as it sits in storage.

Once again, do not limit your preventive maintenance to these items but look after the tools as you would any precision instrument. If these guidelines are followed, it puts the mold maker in a positive position to help answer the "how long will they last" question.

ACKNOWLEDGMENT

Ampco is a registered trademark of Copper and Brass Sales Inc. QC-7 is a registered trademark of ALCOA.

6

AUXILIARY EQUIPMENT

Robert J. Pierce
Ferris State University
Big Rapids, Michigan

6.1 MATERIAL HANDLING

Material handling is an often overlooked aspect of an otherwise well-run operation. What is there to consider? The material is received, it is stored, moved to the machine, the parts are made and the product is stored again until it is shipped to the customer.

As in most other aspects of a manufacturing operation, this is a gross oversimplification of what actually takes place. It is not unusual, however, for the plant operating personnel to be so engrossed in the day-to-day processing operations and the problems of producing quality parts to take the material handling for granted. The fact that the operation is growing, that new materials are being handled and processed, and that volumes of raw materials handled are becoming significantly larger may not be recognized because it is developing so slowly and smoothly, . . . or is it?

As all of these changes take place in an otherwise well-run operation, opportunities for increased efficiency, reduced labor, and actual raw material purchase price savings may be overlooked. Warehouse space may be opened up for use in expanding processing facilities, raw material bags, gaylords, and hoppers may be removed from the shop floor, making other operations, such as tool changing, finished part handling and at machine secondary operations easier and more efficient.

The first consideration should be the material itself. It must be the right material in the right form when it is received. All manufacturing facilities should have receiving procedures that ensure that the material coming in is what it is supposed to be. This can take the form of certified data sheets from the raw material supplier, which guarantee that the material delivered is the material that was ordered, but also that it meets the specifications agreed to at the time of ordering.

If your facility has a quality control laboratory, it may be a good idea to run the raw material delivered through a few QC tests before accepting it. These tests could include visual inspection, screening to check on fines levels, molding test specimens to perform basic physical property tests, such as tensile, impact or heat distortion tests, and running an infrared spectrometric scan to be sure it is the correct material. Exactly which tests are run will depend on how many different materials you receive and how critical mixing them up is to the final product quality and performance. If you are receiving and storing raw material in bulk storage units, *BEFORE* it is put into storage, it is extremely important to be sure that the correct material of the correct quality is being delivered. Otherwise a large amount of material will be contaminated.

Figure 6.1   A multiple silo installation—Peabody TecTank.

6.1.1   Storage

Silos (Fig. 6.1)

The largest potential savings and benefits in material handling may come from the installation of silos for bulk delivery of raw materials. Most material suppliers reduce delivered cost per pound for bulk shipments. These materials may be delivered by bulk over-the-road tank truck or by bulk railroad cars. The bulk trucks deliver about 40,000 pounds of material per shipment and unload themselves. Bulk railroad cars deliver up to 190,000 pounds per load, but are not self-unloading. The receiving facility must provide the system for transferring the material from the bulk railroad car into the silo. How this is done is discussed in the pneumatic conveying system section.

Silos come in many sizes, shapes, materials, and methods of construction and are designed to meet nearly any requirements for special raw material handling.

There are welded silos that can be prefabricated in the silo manufacturer's facilities. These silos are normally finished inside and out in controlled conditions and have all necessary connections already in place. Then, the silos are delivered by special truck to the customer's job site, erected on footings already in place, and are connected to allow loading and unloading. These silos can be transported in diameters up to the maximum allowed by the state highway commissions in the states through which they must be transported from the silo manufacturer to the customer's job site. This is usually up to 12 to 14 feet in diameter (Fig. 6.2).

Figure 6.2   Welded silos—Peabody TecTank.

If the raw material to be stored has a bulk density of 45 pounds per cubic foot, a silo designed to handle two truck loads of material must hold [(40,000Lbs. × 2) × 1.15 safety factor] = 92,000 pounds or 2044 cubic feet of material. With a cross section 12 feet in diameter, or 113.1 square feet, a straight wall silo must be about 18 feet tall.

This calculation does not consider the requirement for a conical silo bottom to allow bottom unloading or the need for clearance under the cone for connecting material handling equipment. Most silo manufacturer's supply tables of the sizes (outside dimensions and height) and capacities of their standard silo designs. The table from one such manufacturer shows a silo of 12 feet in diameter and 24 feet tall for a capacity of 2,000 cubic feet, with a 45° conical bottom. This includes the support skirt which makes the silo taller than the calculation would indicate.

Bolted silos of larger diameter can also be constructed by prefabricating curved and flanged plate sections that are bolted together at the customer's job site. These panels are preformed, sand blasted, coated inside and out, and shipped in racks so that they are not damaged in shipment. Then, they are erected by experienced crews on the foundations provided. This type of silo is easily built in diameters up to 38 feet with capacities over 96,000 cubic feet, or 4,000,000 pounds, if desired (Fig. 6.3).

Silos are manufactured in mild steel (coated, painted, or corrugated and galvanized steel), aluminum, stainless steel, and even bimetallic materials that provide a special surface on the inside (in contact with the raw material) and a more economical material on the outside.

Bulk shipments and silos are usually used for a number of reasons. One is to take advantage of the shipping and packaging savings available from the raw material supplier. Another is to reduce the amount of packaging materials that must be disposed of when empty. Still other reasons are to free up warehouse space for other uses and allow automating of raw material distribution to the processing machines.

Figure 6.3 Bolted silos—Columbian Steel Tank Co.

A number of considerations that should be taken into account before making the decision to install bulk silos, and before writing the specifications to purchase them.

First of all, what are the reasons for doing it? Can the investment be justified economically based on raw material savings, potential savings from automation, other uses of the warehouse space or other advantages? If the answer is positive after a thorough evaluation, then the following points should be considered when writing the specifications for the silos.

How much capacity is required? Consider this question carefully. Think ahead because it is much less expensive to put in an oversized silo today than it is to come back later and take out a smaller silo to replace it with a larger one. Consider the amount of material you will be handling and the space available for silos. It may be better to put in two or more silos if several different raw materials are to be handled. Contamination is a serious problem, and cleaning silos, especially in the winter, can be a big problem. Carefully consider the usage rate for the material. Can a weekend's production be covered with the silo capacity? What about holiday weekends and times when weather delays delivery? A standard size of two bulk loads plus 15% safety factor is a good starting point for bulk truck shipments.

A silo for bulk railroad car shipments may not need to be calculated the same way, but check on the demurrage cost for holding material in the railroad car before installing a silo of smaller size.

The silo manufacturer needs specific information about the plant's site and the raw materials that will be handled in the silo.

The site location information needed includes the earthquake zone the plant is in, what wind loadings must be designed for, and what are the local zoning laws that could affect the silo's height and color.

The silo manufacturer needs to know what materials will be handled in the silo. They will also want to know the bulk density (pound per cubic foot), the angle of repose (the slope of a natural pile of the material), and the flow characteristics of the material.

The flow characteristics of the material are a consideration that the silo customer should also understand and determine what effect they will have on their manufacturing processes.

Materials flow or come out of a silo in one of two ways. They either "funnel flow" or they "plug flow." This is also true of material flowing from any hopper type container, such as the hopper on the infeed of a processing machine or coming from a drying hopper (Figs. 6.4 and 6.5).

Funnel flow means that the material actually flows from the top of the hopper, down through the rest of the material in the hopper and out the bottom spout of the hopper. This is very common with pelletized materials. It means that the ma-

Funnel Flow Occurs When:
1. **Hopper cone walls are too shallow, less than 60°.**
2. **Hopper walls are rough, causing material to drag.**
3. **Non-symmetrical air defuser, material spreader cone.**

Results:
Material flows faster through the center of the hopper and slower along the side walls resulting in uneven drying time.

Figure 6.4   Funnel flow—Conair Group.

terial on the top, the last in the hopper, is the first out of the hopper. This is very undesirable in a drying hopper, but may not be bad in a silo. This type of flow can make contamination problems more difficult to correct, however.

Plug or mass flow on the other hand means that the material flows uniformly from the bottom of the hopper. The first in is the first out. All material in the hopper or silo has the same residence time in storage. This is very favorable in a dryer but may not be necessary in a silo, although it would reduce mixing of the silo contents.

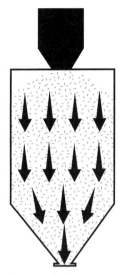

Mass Flow Occurs When:
1. Hopper cone walls are steep, minimum of 60°.
2. Hopper walls are smooth, with no obstructions, shelfs, ledges or protruding welds.
3. Air defuser, material spreader cone is symmetrical.

Results:
All material flows through the hopper at the same rate, resulting in all material being exposed to the drying air for the same period of time.

Figure 6.5   Plug (mass) flow—Conair Group.

The flow characteristics of the material can be modified by the bottom silo hopper cone design. Changing the slope of the bottom cone angle from 45° to 60° changes the flow characteristics of most materials from funnel flow to plug flow. The drawback to doing this is that the silo must be taller to compensate for the reduced volume of the cone bottom resulting from the steeper cone angle. Other design devices can be used to change from funnel flow to plug flow in a silo or hopper also, but they are usually more complicated and expensive.

The silo manufacturer will also want to know how many separate connections are needed and what their sizes and locations must be. Openings are needed for loading and unloading the silo and also for clean-out, venting, and level detection and inventory control measuring devices. The customer should specify the clearance needed between the hopper bottom and the foundations. Enough room must be available to install and maintain the material handling equipment, but there should also be enough room to unload the silo in other ways if there is a contamination problem. There should be room for a manual bottom shutoff valve to isolate the silo from the airvey system when it needs to be worked on. High and low level alarms may also be needed, high alarms to keep from overfilling the silo and wasting valuable raw material and low level alarms to keep from shutting down the operations due to lack of material. Inventory measurement is necessary for cost control, yield calculations, and for reorder timing. There are several methods for inventory control. The simplest method is to drop a plumb bob down from the top of the silo and estimate the amount of material left in the silo by measuring the free volume. In bad weather, it is not a pleasant or safe job to climb the silo to make the measurement. Therefore, products on the market do this measurement automatically and report the measurements in a remote location, such as in a control room. Some of these devices are actually automated plumb bob systems, some use ultrasonic measurements, and some use capacitance probes for the measurements. These systems are suitable for estimating the amount of material left in a silo, but they suffer from the vagaries of varying bulk density and the accuracy of the calculations done with the numbers.

Silos can be mounted on load cells, and strain gages can be mounted on silos skirts to actually measure the weight of material in the silo. These are much more expensive methods for inventory control, but they are also more accurate.

Materials of construction must be specified by the customer. The materials should be selected on the basis of the abrasiveness and chemical reactivity of the raw material and the cost effectiveness of the selection. Silos are difficult and expensive to recoat in the field, so carefully selecting the type and quality of the original coatings is very important. You do not want the coatings coming off in the raw material, and you do not want rust flakes in the raw material either. Proper selection of the gasket material used in assembling the silo and the item connected to it is also important. You do not want red gasket material showing up in your clear molded parts.

The silo customer normally supplies the foundation for the silo based on load data supplied by the silo manufacturer. A local civil engineer familiar with local conditions and regulations should be engaged. It may be a surprise to the customer to find out that the determining design consideration for the foundation is not usually soil loading, but resisting silo tipping during wind storms. The silo manufacturer will also supply the anchor bolt pattern that must be cast into the foundations.

Bulk Bags

If silos are not practical for the current operation, a second method of obtaining savings in packaging, storage and handling costs is to request delivery of the raw material in bulk bags. These bags hold from 1500 to 5000 pounds of material or more. The bags themselves are returnable or single use. They are designed to be hung on a frame above the machine or the material handling equipment and have a bottom unloading valve that feeds the raw material into a hopper by gravity. They are particularly useful for powdered materials that otherwise create a lot of dusting. The frames on which the bags are hung can be equipment with weighing devices to help with inventory control and yield calculations to keep the operators informed when they need to be changed. The bags can be designed to handle materials that must be kept dry or otherwise protected from contamination.

Specially equipped fork trucks must be used to handle the bulk bags. The forks are designed to lift the bags using the loops sewn into the tops of the bags. These loops are used to suspend the bags from their unloading frames.

Cartons

If raw materials are to be received and handled in 1000 pound cartons (gaylords), several considerations help reduce handling and contamination and free up space around processing equipment.

Pneumatic or mechanical conveying systems can be used to transport raw materials from the warehouse (or other remote location) to the processing machines. This procedure keeps the gaylords out of the processing area and reduces the number of material handling personnel needed to supply the machines. Devices, such as gaylord tippers and dumpers, assist in this type of remote material handling operation. Gaylord tippers tip the gaylord from the horizontal and make the raw material flow to the lowest corner of the gaylord to empty the gaylord more completely without manual attention (Fig. 6.6). Many of these tippers are pressure compensated so that, as the gaylord weight decreases, the gaylord is tipped more automatically. Gaylord dumpers actually pick up a full gaylord, lift it the required amount, and dump it upside down into a hopper. The unit should be designed to prevent any foreign material that may adhere to the outside of the gaylord or the pallet from being dumped into the hopper along with the raw material.

Bags

Smaller quantities of powdered and pelletized raw materials are usually supplied in 50 pound or 25 kilogram paper or plastic film bags. These bags are normally palletized with 1500 to 2000 pounds per pallet. The bags may be glued together to keep them on the pallet or they may be shrink wrapped to the pallet.

Bags can be hand unloaded, cut open, and dumped into hoppers for feeding into the processing machines. It is possi-

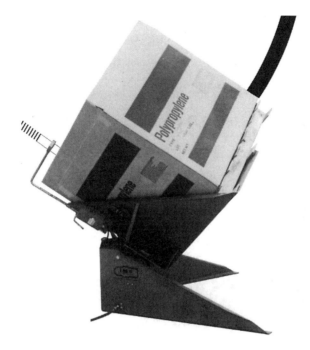

Figure 6.6   Carton tipper—IMS Co.

ble to place a vacuum pneumatic probe directly into an open bag and transfer the raw material into the desired hopper.

The bags must be handled to prevent fragments of the bag material from entering into the processing stream.

There are bag unloading devices on the market. They are often in the form of a hopper covered with expanded metal or grating that allows the operator to place the bag on top of the hopper and cut it open so that it dumps into the hopper without having to handle or support the bag. These units may be equipped with a hood and dust collector system if necessary (Fig. 6.7).

Handling many bags increases the opportunity for contamination and operator injury. It is possible to purchase ro-

Figure 6.7   Bag dumper station—The Young Industries, Inc.

bots that unload bags from pallets and dump them into hoppers to lessen the possibility of operator injury.

6.1.2 Transport

Once the material is delivered, the processing company must somehow distribute and deliver the raw material to the processing machinery. How this is handled depends to a large extent on the form in which it is delivered. An effort should be made, however, to reduce the amount of manual labor and package handling needed to accomplish delivery to the infeed hopper of the processing machine to reduce labor cost and also to reduce contamination.

Pneumatic Conveying

One of the most effective methods for distributing powdered and pelletized raw materials from storage to processing machines is pneumatic conveying. Raw materials may be removed from their containers, conveyed in enclosed lines, and delivered to the infeed hopper on the processing equipment without exposure or handling by human hands. This process can be done cleanly and with minimum degradation of the raw materials being handled. The various methods which can be used are listed following.

Dilute Phase

This type of system uses relatively large amounts of low pressure air to convey the suspended raw material at relatively high speeds from the source to the delivery point. It is called dilute phase because there is space between the individual particles as they are transported through the conveying system.

Vacuum Systems. A relatively easy method for conveying materials is done by creating a vacuum at the receiving end or hopper that pulls the raw material into the conveying line with the air that is pulled in to relieve the vacuum. A specially designed "wand" has a center pipe that conveys the raw material by air flow through it which is surrounded by an annular opening that allows air to flow down to the pickup point. In another design, the raw material is loaded into a hopper that has a tee type outlet. The raw material flows in from the top, and the air flows in from one side (usually through a filter) so that the air velocity picks up the raw material and conveys it to its destination.

Metal pipe is usually used as the conduit through which the air and raw material flow. The conduit could also be plastic, ceramic, or glass pipe. This conduit is routed from the sup-

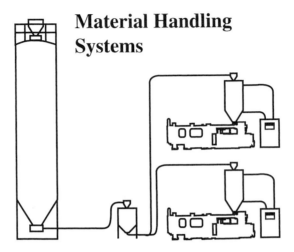

Material Handling Systems

Figure 6.8   A typical pneumatic conveying system—Conair Group.

ply to the delivery point by using elbows, couplings, and straight runs of pipe (Fig. 6.8).

Flexible hoses may be used at any point in the system where there must be options for the flow path. This could be at the pickup end where several source locations might be used, along the flow path where different routes may be needed, or at the discharge end where several destinations might be needed.

At the discharge end, several methods are used to separate the raw material from the air stream. One common method is to allow the raw material to be collected in a hopper. This allows the air velocity to slow down enough so that it no longer conveys the raw material and drops it in the hopper as the air exits from the top of the hopper (Fig. 6.9). This hopper is usually equipped with a bottom valve that is held shut by the vacuum in the hopper during conveying. A level switch or timer controls the vacuum system, and when the hopper has a full load, the vacuum system is shut off. The weight of

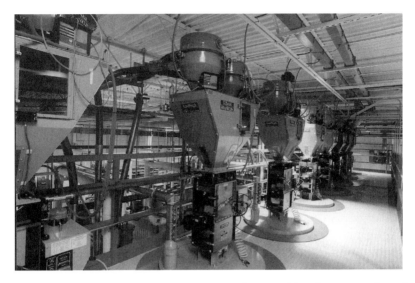

Figure 6.9 Receiving hopper on top of a machine hopper—Conair Group.

the material in the hopper opens the bottom valve and dumps the material into the infeed hopper of the processing machine. This cycle can be repeated as often as needed to maintain the desired operating level in the infeed hopper.

This process ranges in size and complexity from the simple "hopper loader" designed to be self-contained and load material from a source located close to the processing machine to centrally located, computer-controlled systems that supply several machines from sources as far away as 500 feet from the multiple machines supplied.

It is possible to design a vacuum pneumatic system that does not have an interrupted service cycle. In this case, a rotary or star valve (also called an air lock) replaces the vacuum-operated bottom valve on the receiving hopper. This valve has a rotating vane with multiple pockets that fill with raw material as the pocket rotates past the bottom discharge of the receiving hopper and dumps the material into the infeed hopper as it passes its inlet. The number and size of the pockets in the rotor are designed so that there is never a clear path for air flow from the infeed hopper into the receiving hopper, thereby maintaining the vacuum seal in the receiving hopper (Fig. 6.10).

Pressure Systems. In a pressure dilute-phase conveyor system, air flow is supplied at the infeed end of the system. This has the advantage of allowing higher pressures (high energy) than achieved by a vacuum system (maximum of atmospheric pressure, 14.7 psi) which translates into longer possible runs. On the other hand, there are also some disadvantages. As energy is used in conveying the raw material through the system, the internal pressure is reduced, and the air expands. This expansion results in accelerating of the raw

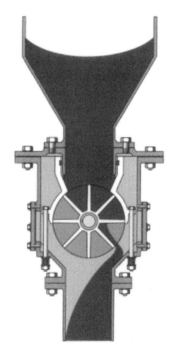

Figure 6.10 Rotary star valve—Detroit Stoker Co.

material as it moves through the system. The complete system must be designed so that the velocity is not so high that the raw material is abraded and fines are formed. Secondly, because the conveying system is under pressure, there must be air locks at each raw material pickup point to keep the air pressure from just blowing back up into the supply hopper and not conveying the material down the pipe. The more pickup points used, the more airlocks are needed. Rotary valves can be a large maintenance problem, because of wear and breakdown and because of the damage they can do to the raw material by shearing the particles.

When constructing a vacuum or a pressure pneumatic conveying system, a number of details should be considered.

First, the pickup velocity, the air velocity needed to pick up the raw material and start it flowing with the air stream, must be known. This information can usually be obtained from the raw material supplier. It is frequently in the neighborhood of 4500 feet per minute for pelletized materials. The maximum flow velocity before significant raw material degradation takes place should also be obtained from the raw material supplier. This depends on pellet geometry and how brittle the material is. Then, the system must be designed to maintain the velocity in the system between these two values.

Frequently, the straight lines for conveying are made of aluminum pipe or tubing. If the material is very abrasive, other materials such as steel, stainless steel, ceramic or glass may be needed. Once again, check with the raw material supplier for its recommendation.

Elbows that carry the raw material should be made of stainless steel. Almost any material will wear holes in aluminum elbows in a fairly short time. The elbows should have a radius of curvature that is 12 times the diameter of the pipe or tubing. In other words, an elbow made of 3 in. diameter pipe should have a radius of curvature of 36 in. Each elbow must have a straight section on each end to connect it to the straight sections. The inside radius of the elbows must be wrinkle-free. There is a tendency for the metal on the inside of the bend radius to wrinkle as the pipe or tubing is bent to form the elbow. Elbows with these wrinkles should not be accepted.

The previous discussion applies to rigid pelletized raw materials. If elastomers or softer types of raw materials are to be conveyed, the design parameters are somewhat different. First the inside of the straight runs of pipe or tubing must be treated to stop the formation of "angle hairs" or streamers that are caused by the raw material sticking to the inside surface and being stretched into long, fine strands that will later plug up the system downstream. There are several methods for this treatment and your material supplier can recommend the correct method for its material.

Elbows for elastomeric, soft, very abrasive or friable materials should be designed differently from those for rigid pelletized materials. Rather than a bent elbow, a right angle corner needs to be formed with a pocket at the end of the straight run coming up to the elbow that can fill up with the raw material and form an impact surface of the raw material itself. The material being conveyed can run straight into the material in this pocket and bounce off and on around the elbow without abrading the elbow itself and usually without damage to the raw material either. These elbows can take the form of a standard pipe or tubing tee of the same size as the straight runs. They need to be installed so that the blank extension of the tee is downstream of the material flow coming from the straight run.

In some cases it may be wise to use an elbow that is specially designed for this type of installation. It is usually made of cast metal and may have a replaceable end cover that is easily replaced without disassembling the system if worn out (Fig. 6.11).

The couplings used are usually designed to slip on the outside of the pipe or tubing and clamp in place. They must have the correct diameter to ensure correct alignment. If the pipe or tubing diameter is more than 3 in., it should have four clamping blots. Usually these couplings have a rubber liner to give some flexibility to the joint. There should also be a

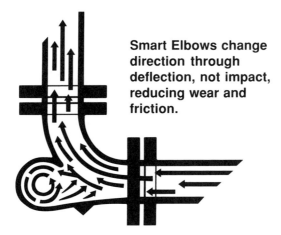

Smart Elbows change direction through deflection, not impact, reducing wear and friction.

Figure 6.11 Wear resistant elbow—HammerTek Corp.

grounding strap to provide electrical continuity from pipe section to pipe section. When conveying plastic materials, especially when they are dried, very high static electrical charges can be stored up in the system unless it is thoroughly grounded. These charges can cause problems ranging from uncomfortable shocks and fires to separation of color concentrates from natural resins, resulting in color problems and concentration of fines. The whole system should be grounded.

When laying out the routing for the pneumatic conveying system be sure to take clean-out into consideration. Accessibility is the first consideration. Do not run conveying lines where they cannot be reached, at least with scaffolding or a "cherry picker." Remember that an airvey line full of material is going to be heavy to handle. The slip couplings are used to make it easier to remove a section of airvey pipe or tubing without disturbing the rest of the system.

Also place the major pieces of the system, such as blowers, valves, and receivers for easy maintenance. Otherwise, they will not be serviced until there is a major problem.

Many types of air moving devices are used in pneumatic conveying systems. They include venturi systems that use compressed air to cause the vacuum that moves the material

for short distances from supply container to machine hopper
and positive displacement rotary compressors that develop
high vacuums or pressures and move large amounts of air
and raw material over long distances. Centrifugal fans are
used for low pressure and vacuum installations, and turbine
blowers are used for intermediate distances. Each of these
units has advantages and disadvantages. The positive displacement units provide the most pressure with the most efficiency (Fig. 6.12). Their output is not affected by changing
conditions downstream in the system, but they blow a
plugged system apart if pressure relief is not provided. These

Figure 6.12 A Roots blower package for pneumatic conveying—
Roots Division, Dresser Corp.

units cannot stand to have solids pass through them. They must be provided with filter protection to prevent raw material from entering them. Positive displacement blowers are often very noisy, need mufflers and sound enclosures, or must be located where the noise is not a problem for personnel. Turbines pass some raw material, but often have aluminum blades that are quite fragile and also need to be protected. Centrifugal fans have large clearances and are usually made of steel. This means that they can pass raw material without significant damage, at least for a while. They will damage the raw material, however. Centrifugal fan performance depends strongly on changes in the flow and back pressure in the system. Relatively minor changes in these parameters can effectively shut the system down. Therefore, careful consideration should be given to the exact type of air moving device selected for the specific application.

Dense Phase
Systems are designed for powdered materials and those that are easily degraded in dilute-phase conveying. These systems usually consist of a pressurizable batch tank for the material to be conveyed, connection pipes and elbows designed for high pressure, and a receiving tank or hopper at the final destination. In these systems, the batch tank is loaded with the material to be conveyed, closed up, and the batch tank is pressurized until the raw material is pushed in "slugs" from the tank through the pipe to the receiving hopper. There may be booster jets along the pipe to keep the material moving in the system.

One advantage of these systems is that they are totally enclosed. They limit the opportunity for contamination and control dusty emissions. The angle hair or streamer problem is also normally eliminated by these systems.

Mechanical
Mechanical conveyor options are also available. Generally they fall into two types, the belt conveyor and the auger conveyor.

Belts.   Belt conveyors use a flexible belt, usually rubber, but it could be thin metal if needed for chemical or heat resistance. The belt runs over a series of rollers with a drum pulley at each end. The raw material is dumped on the belt at the source end and is dumped off the belt at the receiving end. Usually these belt conveyors are relatively open and allow contamination. They can be designed with a weight section in them and then continuously weigh the amount of material passing over the belt. This is discussed further in the weigh feeder section.

Belt conveyors can develop problems with belt tracking (keeping the belt running straight and on the rollers) and with raw materials that stick on the belt and travel back to the source on the belt or build up on the drum pulleys. Frequently, brushes are used to clean the belt at the discharge end. Another option with a belt conveyor is to pass the raw material over or past a magnet that removes any tramp ferrous metal contained in it.

Augers.   Auger conveyors use a spiral that carries the raw material along as it rotates within a tubular housing. It works in exactly the same way as a grain auger conveyor.

The spiral auger may be a solid flight "screw" or a spiral coil that is open down its center. The solid screws can be used only in straight runs because they cannot bend. Spiral coils can be used in a bent tube, as long as the bends are not too sharp.

Both systems are usually used for relatively short runs.

6.1.3 Feeders

Most feeders are used to add color concentrate to natural polymers to produce the desired color in the final parts, but feeders may be used to add nearly any additive. Flame retardants, antioxidants, ultraviolet light stabilizers, impact modifiers, and many more additives can be handled as concentrates or as powders. There are two classes of additive feeders, volumetric and loss-of-weight or gravimetric.

Volumetric

The feeders in this class work by feeding a constant or controlled volume of an additive into the processing stream. They do not weigh the material but rely on a constant bulk density for the additive being added. Bulk density can vary, depending on packing, clumping, and the level of the additive in the supply hopper, but in many cases it either does not change significantly or at least not enough to be a problem. The main advantages of volumetric feeders are low original cost and ease of maintenance. Most volumetric feeders are straightforward mechanical units that can be maintained and repaired by a mechanic or millwright. The associated electronics are usually fairly simple. They can, however, be tied into a process control unit that controls their feed rate based on the production rate of the process, but they still control by volume.

Belts

These units look like short belt conveyors. The main difference is that they have variable speed drives so that the rate can be adjusted to give the volumetric rate required. Most belt feeder manufacturers have testing or demonstration laboratories where the exact materials that you wish to feed can be demonstrated and tested to help select the correct belt feeding system for your application.

Augers (Fig. 6.13)

Auger feeders work on the same principles as auger conveyors but have three important components. There is the infeed hopper, which may or may not have an agitating device in it, depending on the characteristics of the additive that will be fed from the hopper. If the material tends to bridge, rat hole, or clump together, an agitator is required. The second component, the one that actually does the feeding, is the auger system. This may have one or two auger screws. They take the form of coil spirals or solid screws of many different shapes or cross sections. Each design is for a specific type of additive. It may be desirable to have several different sets of augers for any one feeder so that it can be used effectively with different

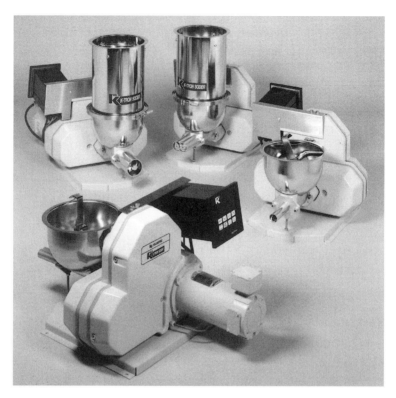

Figure 6.13   Auger Feeders—K-Tron Corp.

additives. The third component of the system is the variable
speed drive unit. With this unit the actual feeding volume is
controlled by changing the rotating speed of the augers. These
controls can usually be tied electrically into a process control
computer to change the speed automatically when the overall
processing speed varies. As with the belt feeders described
previously, most feeder manufacturers have testing laboratories where the exact products you wish to handle can be tested and demonstrated so that the correct components can be
selected.

Loss of Weight

A further development of volumetric feeders are loss-of-weight or gravimetric feeders. These units have a weighing device incorporated into them that allows the feeder to exactly weigh the material fed into the process. The advantage is much more accurate addition of the material to the process. The more accurate addition rate saves money by ensuring that enough additive is added and by reducing the possibility of degrading final part performance because either too much (poor performance and added cost) or too little (poor performance or color, not enough impact,) additive is added. The disadvantage of these units is that they are usually several times more costly than a similar volumetric feeder and they take a more skilled technician to maintain and repair.

Belts (Fig. 6.14)

This type of unit is physically similar to a volumetric belt feeder. The only differences are in the electronic control sys-

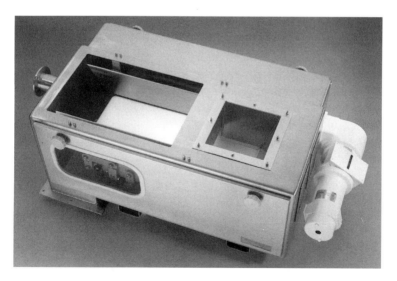

Figure 6.14 Weigh belt feeder—K-Tron Corp.

tem and the weight deck that is part of the belt conveyor system. The electronics are set up to constantly calculate the rate of material passing over the belt based on the weight on the scale and the rate of the belt travel. These units are normally used where a larger amount of additive must be handled and the process is continuous. A regular schedule of calibration checks should be established and followed to be sure that the weigh system is correct at all times. Experience will dictate the exact schedule needed. Most manufacturers have testing laboratories for selecting the correct system for your application.

Augers (Fig. 6.15)
Loss-of-weight auger feeders look very similar to volumetric feeders. The main difference is in the electronics and the infeed hopper (or the whole unit) which is mounted on a weighing device. The electronics use the data from the weighing device against time to calculate the addition rate to the system. Then, the result of the calculation is used to adjust the speed of the augers to control the addition rate. These units are more easily used in processes that are intermittent, such as injection blow molding. In injection blow molding, the feeder runs only when the plasticating screw is running. Most manufacturers have testing and demonstration laboratories to assist in selecting the correct system for your application.

6.1.4 Dryers

More engineering polymers are being applied in the market place, and most of these polymers need to be dried. Some of the old standby polymers, like ABS, PET, and polycarbonate, also need to be dried to produce high-quality finished parts. Some materials also process easier in the plasticating screw if they are preheated in a dryer before being feed into the screw. Therefore, polymer dryers are becoming more important to a successful processing operation.

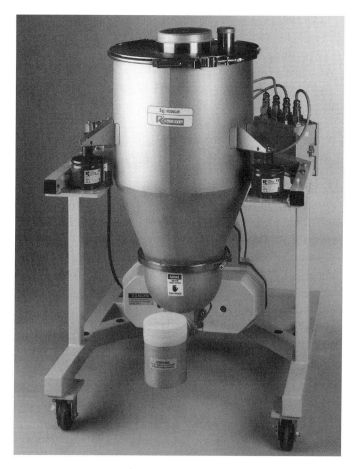

Figure 6.15   Loss-of-weight feeder—K-Tron Corp.

Basics

The following four parameters must be present for any polymer drying system to work properly.

Air Flow

If there is no air flow through the drying chamber, the moisture released from the polymer will not leave. It will still be

present when the material is fed into the process. Secondly, hot air flow is usually the means to heat the polymer to the temperature needed to release the moisture. An air flow rate of at least 1.0 fpm (feet per minute) across the cross-sectional area of the drying hopper must be achieved. In other words, if the hopper is 3 feet in diameter, the cross sectional area is 7.07 sq. ft., and the flow rate should be at least 7 cubic feet per minute through the dryer hoses. More air flow may be needed to properly heat the polymer.

Temperature

Most polymers release moisture much more rapidly at an elevated temperature. Some will not release it at all unless they are above a certain temperature. On the other hand, if the polymer is heated to too high a temperature, it softens and sticks together, called clumps or "hot balls." Therefore, temperature control in a dryer is very important to the performance of the dryer. Follow the recommendation of the polymer manufacturer for the temperature that the material must maintain for proper drying.

Time

Once the polymer is at the specified temperature, it must be kept at that temperature long enough to drive the moisture out of the center of the pellets or granules. The polymer manufacturer will also specify the amount of time the material must be at temperature to ensure complete drying.

Dew Point

The last, but not the least important, parameter is the dew point or relative humidity that is specified by the polymer manufacturer. The correct equilibrium of temperature, air flow, time, and the absence of moisture in the surrounding atmosphere must be established to insure complete drying. Some polymers give up moisture relatively easily and can be dried in an atmosphere that is at a dew point of $-3°F$ ($-19°C$). Other materials must be in an atmosphere that has a dew point as low as $-20°F$ ($-29°C$), or even lower. This parameter is so important that it is common practice to equip dryers with built in dew point meters. They can control the regeneration of the desiccant to ensure that the dew point does not rise above the set point.

Types of Dryers

There are several configurations for drying polymers. They vary mostly in sophistication and cost. The amount of sophistication is determined by the types of polymers that have to be handled and the demands for drying them. The cost is directly related to the sophistication of the system.

Hot Air (Oven) (Fig. 6.16)

Some polymers can be dried by putting them in shallow pans in a drying oven and holding them at a specified temperature for a specified time. These drying ovens normally have circulating fans but do not have desiccating systems. It is possible to attach desiccating units to these oven if need be, but it is

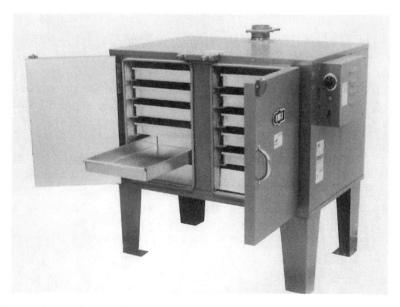

Figure 6.16   Pan drying oven—IMS Co.

usually more efficient to purchase a desiccant dryer to handle those polymers that require it. Oven dryers are useful for short runs, prototype shops, or quality control labs where small amounts of material are dried and time is not an important variable.

Refrigerated

These units are not seen very much any more, but they may still be in use in some shops. A refrigeration unit is used to drop the dew point of the air circulated in the system by cooling it and condensing the moisture. The lowest dew point these units can reach is in the neighborhood of freezing because the condensate freezes and plugs up the drains if colder temperatures are reached. The other inefficiency of these units is that the air must first be cooled to remove excess moisture and then reheated to the drying temperature specified by the polymer manufacturer. These units are not in general use today.

Desiccant Systems (Fig. 6.17)

Desiccant systems are by far the most common systems in use today for drying polymers. Several unit operations are included in the desiccant system. These include the air mover, the air heater and the heart of the unit, the desiccant or moisture removal unit. Desiccant is a manufactured zeolite material that has a porous surface with holes or cavities sized for water molecules. These materials also have an affinity for water, i.e., they attract moisture. In other words, these materials selectively remove water or moisture molecules from an air stream that is passed through them. This is a physical process, the desiccant eventually becomes saturated, and all the cavities fill up. When this happens, the desiccant bed must be "regenerated." This means that the bed must be heated to a temperature to drive off the moisture from the bed. Regeneration takes a temperature of 425°F (218°C) or higher. It also takes some time. Naturally, the desiccant material is also heated and tends to hold this heat. If the desiccant bed were put back in service to dry polymer immediately after regeneration, the air would be overheated and the polymer would melt and con-

Figure 6.17   Desiccant dryer installation—Conair Group.

geal. Therefore, the bed must be cooled before it is put back in service. As a result, most desiccant drying systems have at least two desiccant beds so that one is in service while the other is being regenerated and cooled.

There are several variations on this system. One company uses a rotating unit with several smaller desiccant units so that one is in service, one is regenerating, one is cooling, and at least one is ready to go into service when needed (Fig. 6.18).

Another variation has desiccant adhered to the surface of a honeycomb wheel that is constantly rotating so that the material in service passes through that zone and moves on to the regeneration zone, then to the cooling zone, and back into the

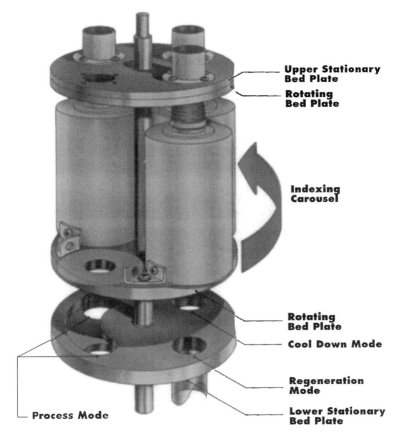

Figure 6.18 Desiccant carousel—Conair Group.

service zone. The main claim for this unit is that it provides a constant dew point at all times.

There are also some variations on the other unit operations with different desiccant systems.

Air heating units may be placed in the desiccant unit remote from the drying hopper or there may be "booster" heaters at the drying hopper, or both. The second and third systems have some advantages. First, the same desiccant unit can be used to supply more than one drying hopper, each of which can have different polymers in them that require different drying temperatures. Secondly, the heater at the hopper insures that the air entering the hopper is at the correct temperature. With remote heaters, it is possible for the air to cool significantly before it reaches the hopper.

Some polymers require relatively high drying temperatures. The desiccant requires relatively cool air passing through it to remove moisture efficiently. This often requires an "aftercooler" between the drying hopper and the desiccant unit in the return air line. The air returning to the desiccant unit should be 130°F (54°C) or less for full efficiency. If the air gets above 150°F (65°C), the desiccant cannot remove moisture effectively. Most dryer manufacturers have aftercoolers designed to work with their dryers. These aftercoolers can be either of the air-cooled or water-cooled variety.

The design of drying hoppers is based on three important considerations. First, they should be insulated so that the air flowing up the inside surface of the hopper wall is not cooled below the specified temperature. Insulating the hopper also reduces heat losses and improves operating efficiency. Secondly, the hopper must have some sort of air diffuser that ensures that the air flow does not "channel" up through the polymer pellet bed, so that all the material in the hopper is exposed to the same amount of air flow. Thirdly, the hopper must be designed for plug flow of the polymer pellets. This is necessary to insure that all pellets passing through the hopper receive the same exposure time for drying. It is common in hoppers for pellets to "funnel flow" (see Fig. 6.4) or for the pellets on top to actually flow down through the center of the hopper. If this happens, the last material in is the first out and does not stay in the hopper long enough to dry properly.

Lastly, it is important to understand that desiccant has a useful service life. It degrades over time. It begins to break down to become a powder, and blows through the system. It can also "load up" with impurities from the polymer and lose efficiency because the pores become plugged. The desiccant supplier can test the desiccant to see how efficient it is and recommend whether or not it is time to change it. If the time between regeneration cycles decreases, it is time to check the desiccant. Desiccant can be expected to last two to three years in "normal" service.

Two classes of desiccant are in common service as polymer pellet drying desiccants, 4A and 13X. These numbers are the pore sizes of the desiccant. The 13X material has bigger pores and higher moisture capacity, but it adsorbs other molecules beside water. The 4A has smaller pores, lower capacity, and adsorbs only water molecules. This is important if your facility is running materials that have a lot of plasticizers or other additives that are released during drying. If 13X is used with these polymer pellets, the other additives are adsorbed into the desiccant and are often baked in during regeneration, reducing the efficiency of the desiccant. An aftercooler before the desiccant bed can condense most of these additives before they reach the desiccant. The 4A desiccant does not allow the additives to enter the pores and plug them.

Vacuum

Another way of removing moisture released from a polymer is by vacuum. This approach removes the need for air flow and for dew point control, but it causes some other problems. First, air flow is the usual method for transmitting heat energy to the polymer. In a vacuum dryer, by definition, there is no air flow to do this. Also by definition, there is no air for convective heating. The polymer pellets must be heated by radiant energy. This can cause uneven heating and must be watched so that all the pellets are heated but that some areas don't get overheated. One method for doing this is to agitate the pellet bed with a mixing agitator or by tumbling the pellets in a drying drum. The drying drum also allows for some heating by contact with the heated drum surface.

One other advantage of a vacuum dryer is that the material may have less color development because there is less or no oxygen present during the drying process.

Maintenance

There is a tendency for dryers to be installed and forgotten, at least until a problem develops. Dryers are like any other piece of processing equipment and should receive attention regularly to detect, prevent, and correct minor problems before they become major problems.

All dryers have filters in their systems. These filters should be checked and cleaned regularly because they can significantly reduce the air flow rate through the system, especially in systems that use centrifugal fans to move the air.

Air flow can also be affected by moving a dryer and not checking the rotational direction of the blower motor.

The actual dew point achieved should also be checked regularly. If the desired dew point is not being achieved, there are several possible causes. Besides the obvious one of old and plugged desiccant, the dryer regeneration heater may be burned out, the filters in the regeneration system may be plugged, or the regeneration system may not actually be switching from one desiccant unit to the other. There may be leaks in the system that allow outside air to infiltrate the dryer, overloading the desiccant unit. Wet outside air could also be introduced into the system by a faulty material handling system.

Specification

When contacting a dryer manufacture for a quotation on new drying equipment, there are several pieces of data that you should have available: the range of polymers that will be dried; the temperatures, dew points, and times required for each polymer; the production rates expected; how you expect to mount the drying unit; and any special material of construction requirements you have. It is recommended that a unit be purchased that has built-in dew point monitoring and regeneration control and that an aftercooler be selected. Dryer units are available that are self-unloading with desiccated air. This can be important if very hygroscopic polymers are processed. These units have a very small receiving hopper on the infeed port of the processing equipment and convey from the dryer to the small hopper with dry air so that the polymer is not exposed to wet ambient air between the dryer and the processing machine.

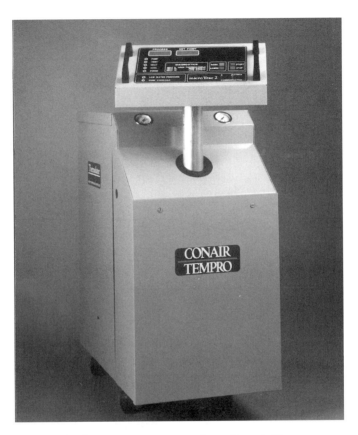

Figure 6.19 Mold temperature controller—Conair Group.

6.1.5 Mold Temperature Controllers (Fig. 6.19)

In many cases, it is sufficient to use well water, city water, or
cooling tower water directly to cool molds. In other cases, the
mold temperature must be controlled within a specified temperature range, whether it is colder than ambient or hotter. In
these cases, a mold temperature control unit is needed.

Mold temperature control units are designed specifically
for cooling, for heating or for a combination of heating and
cooling. In any case, the unit must perform several functions.
First, it must have a circulating system with a flow rate that

yields turbulent flow in the mold fluid channels. If turbulent flow is not established, the temperature control will not be reliable or efficient because laminar flow does not allow all of the fluid to come in contact with the heat transfer surfaces. The flow rate must also be high enough to maintain a temperature difference between supply and return of no more than 4°F (2°C).

The Reynolds number (N_f) may be used to determine the amount or quantity of liquid flow that is needed to yield turbulent flow in the cooling channels:

$$N_f = \frac{[(7740)VD]}{n} \text{ or } \frac{[(3160)Q]}{Dn}$$

where V = fluid velocity in ft/sec
D = diameter of cooling passage in in.
n = kinematic viscosity in centistokes
Q = coolant flow rate in gpm

Note that the kinematic viscosity changes with temperature. If the Reynolds number is above 5000, the flow is turbulent. Be sure that there is enough flow in all channels if parallel flow headers are used, otherwise some channels will have good cooling and others will not!

Secondly, the unit must have heaters and/or refrigeration units that maintain the set temperature under the heat load from the mold. This load (Q) can be estimated by multiplying the part weight in pounds (W) by the specific heat (C_p) for the polymer in BTU/lb.-°F by the difference between melt temperature (T_{mt}) and mold temperature (T_{md}) in °F and then dividing the result by the cooling cycle time (t) in hours:

$$Q = \frac{[(C_p)(W)(T_{mt} - T_{md})]}{t}$$

This calculation yields an estimated heat load for the unit. Contact the polymer manufacturer for the specific heat value for the polymer to be handled. The unit must also provide makeup water and a drain connection.

6.1.6 Cooling Towers and Heat Exchangers

To reduce water usage and the resulting waste water disposal costs and problems, it may be worthwhile to install a cooling tower or heat exchanger (air cooler) to supply a closed-loop cooling water supply system for your operations.

Whichever system is chosen, some common elements of the systems are necessary. First, there must be a closed circulation system with enough pressure to provide the flow rate required for the cooling load present. This system will have pumps, filters, and pressure regulators. More than one pump is necessary to ensure that the system continues to operate if there is a problem with one pump. Multiple pumps may also be desirable if there are regular or cyclic variations in water rate demand. At least two parallel filter systems should be available so that one is in service while the other is being cleaned or replaced. These filter systems should be equipped with a differential pressure gage so that the pressure drop across the filter can be used to indicate a plugged filter.

Both systems also require a chemical balance monitoring and additive system. In most cases, the systems can also dump water from the system to the drain and add fresh water to help keep chemical balance in the system. These systems are used to reduce rusting and scale build up. In addition, they control organic growth and foaming.

Cooling towers consist of a water circulating pump, a tower with some sort of "packing" or expanded material over which the water runs to create a large surface area and, usually, fans that circulate air through the packing to cool the water. Warm water is pumped up from a sump under the tower, sprayed or distributed over the packing, and allowed to trickle down through the packing back to the sump while cooling air is drawn up through the tower packing. Cooling towers can be identified by a plume or cloud of water vapor that rises from the top of the tower. Some of the cooling is done by heat exchange with the water, but much of it is done by converting some of the water to vapor (which take about 1100 BTUs/lb of water). This evaporation is what produces the cloud-like plume from the unit. Of course, this means that makeup water must be added to replace the evaporated water. This system yields water temperatures about 10°F (5.6°C) below ambient air temperature. The systems normally function during winter at temperatures below freezing because the water is warm enough to keep the tower from freezing. The fans may be shut down if the air temperatures get too cold.

Heat exchangers or air coolers may not have a sump but may be part of a truly closed system for circulating the water. The heat exchanger takes the form of a large radiator (similar to an automobile radiator), usually installed horizontally, with single- or multispeed fans to draw cooling air through the radiators. These systems do not require as much makeup water, and they do not emit the water vapor cloud emitted by a cooling tower. On the other hand, they cannot cool the water below the ambient temperature and usually run several degrees above ambient air temperature. Care must be taken when operating these units in winter to keep them from freezing. Antifreeze may be used in the water to control this problem.

If cooling water at a lower temperature is needed for part of the process, a chiller or refrigeration unit can be used in series with a cooling tower or air cooler to produce the needed temperature.

6.1.7 Processing Environment Control

When the humidity and the temperature are both high, it may not be possible to run the mold temperature as cool as desirable because moisture from the atmosphere will condense on the mold. There may be other times when parts have to be left in the mold longer than desired. They must have extra cooling before being removed because the atmospheric temperature is too high and they may distort because they do not cool quickly enough outside the mold.

When conditions like these prevail and if your facility is not air conditioned, it may be advantageous to use a local environmental control unit. Large and small, portable and stationary dehumidifying and air conditioning units are designed specifically for this type of service. These units are located near the processing equipment and blow dehumidified and cooled air directly on the area that needs it. They may or may not include an enclosure to contain their output and localize the effect.

These units are usually electrically operated, but they may be either air- or water-cooled. A refrigeration unit needs cooling for its coolant condenser, so the unit must be adaptable to the cooling service available in your facility. Water-cooled units are more compact, take up less floor space, and release less heat to the work area, but they need a cooling water supply and a drain.

6.1.8 Rework Grinders (Fig. 6.20)

There are several approaches to handling rework. One is to collect it and grind it up when you have extra labor time available. Another is to grind it immediately and feed it back into the system right now. There may be good reasons for each approach. The first method is probably used when there are a number of off grade parts that come from a process to which they cannot be fed back immediately, or they were recognized as off grade after the process was switched to another product. The second method is used in processes that are running at the time and where a predictable amount of flash, runners, and trim is handled. Another advantage of the second method is related to drying. If the material being processed requires drying, it is better to reprocess it quickly, before it has time to readsorb moisture from the atmosphere. In any case, rework grinders are needed in almost any plastics processing operation.

The type and size of grinder needed or used depends on the way it is to be used. If parts and scrap are saved up and ground all at once, a larger grinder is needed. If the runners, sprues, and flash are to be ground press side and fed right

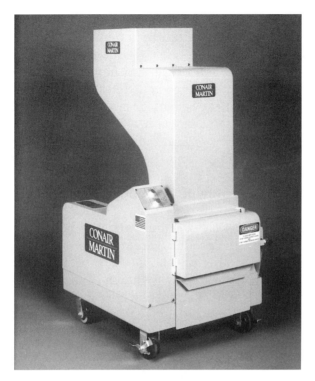

Figure 6.20   Scrap grinder—Conair Group.

back into the process, a smaller grinder is needed because of floor space requirements and the need to save power to drive the grinder.

Most grinders consist of an infeed chute, usually protected so that operators cannot get their hands into the cutters and to keep materials from being flung back out the chute, the cutter system itself with drive, a perforated screen with holes sized to let only particles of the proper size fall through, and a collection hopper of some sort that also has safety protection.

All grinders have to be equipped with safety switches, brakes, and interlocks that will stop or keep the cutters from running if the covers or guards are opened. Grinders are VERY dangerous pieces of equipment.

When selecting a grinder, it is a good idea to take advantage of the grinder manufacturer's testing laboratory to test grind the products you expect to put through the grinder. Does the proposed grinder produce the desired particle without too many fines or oversized particles? Does it do it quietly? Or can sound enclosures be added to control the sound level to required limits? How easily is it opened, cleaned, and maintained? Remember that cross contamination of regrind can be a big problem. Is sharpening the cutters easy, or can the cutter blades be replaced quickly and easily? Are spare parts kept on hand and available at all times from the manufacturer?

The correct particle size and particle size distribution are very important in the product coming from a grinder. The size and shape affect drying and hopper flow characteristics. The larger the particles, the slower they will dry. The shape and size change the angle of repose, which may lead to "rat holing," a condition where the angle of repose is 90°, or the material does not flow at all in a hopper. The bulk density may change from virgin pellets, increasing or decreasing, which certainly changes the calibration of any volumetric feeders in use. Too many fines can accumulate and cause color problems, splay problems, and can also cause high local concentrations of additives, such as lubricants, plasticizers, and other modifiers that will change the processing and final properties of your product.

6.1.8 Miscellaneous

Other accessories that help in making a smooth operation and reducing processing problems are discussed here.

Magnets (Fig. 6.21)

Magnets are used in the raw material supply system to remove tramp ferrous metal. These magnets can be placed in the infeed hopper or in a spacer box between the infeed hop-

Figure 6.21   Drawer magnet—Eriez Co.

per and the throat of the processing machine. In the second case, the magnets are usually installed in drawer-like setup that allows the magnets to be pulled out for cleaning without stopping the process.

Fines Separators (Fig. 6.22)

Pneumatic conveying systems and some other materials handling equipment and grinders may generate fines. These fines can be a processing problem because they may contain higher than normal levels of additives, such as mold release agents and color. They may also contribute to clumping, blocking, and poor feeding of the raw material into the processing machines. As a result, it may be desirable to remove the fines from the raw material stream.

Depending on the nature and volume of fines, there are several options for removing them. The most common method is to use a screener. The raw material is passed over a vibrating or rotating screener. The fines drop through the fine mesh screen and the normal pellets pass over it.

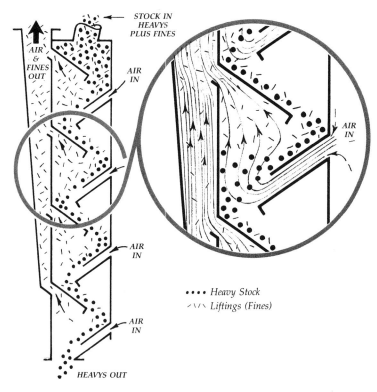

Figure 6.22 Fines separator—Kice Industries Inc.

Units can be incorporated into a pneumatic conveying system that allow the pellets to fall by gravity through a baffled tower or column while a carefully controlled air current passes up through the same column at a rate to convey the fines up and out of the pellet stream (see Fig. 6.22).

7

THE USE OF REGRIND

Dan Weissmann*

Schmalbach-Lubeca Plastic Containers USA, Inc.
Manchester, Michigan

The use of regrind in extrusion blow molding, whether in bottle production or large industrial part blow molding, is usually a must. The design of the extrusion blow molding system is such that some amount of plastics end up adjoining the part and require trimming during the part finishing process.

The main recycled parts in bottles are the moil and the tail. Both are formed when the mold closes on the parison. One end of the parison is pinched together and forms the bottom of the bottle, while at the other end, adjoining the bottle opening, a moil is formed to provide better control of the blowing of the finish where dimensional compliance of the formed

Current affiliation: Consultant, Simsbury, Connecticut.

part is very important to the successful application of caps. The moil could also be used for removing the blown bottle from the mold and placing it in the trimmer, where it helps to stabilize the bottle during trimming.

There are other areas in bottles where trimming must be done. One is the pinched-off area inside a blown handle. Another in large bottles is where the parison is larger than the finish opening of the mold and pinching of the parison occurs along the finish and into the shoulders of the bottle. Typical bottles as molded, with the pinched areas before and after separation, are shown in Fig. 7.1. Industrial blow molding involves much more complicated shapes and designs and parison pinching may take place almost around the entire periphery of the part, resulting in large amount of trim (Fig. 7.2). Other operations, such as hole punching, add to the material to be recycled.

Depending on the size and the shape of the part, the trim can be rather large compared to the material left in the part. Economic considerations-the cost of the material used, make reusing this material a must. Reuse into the same production run normally returns the highest value for the material because it is substituting pound for pound of virgin plastics.

The trimmed off material together with start-up scrap and occasional rejected parts due to noncompliance with quality standards are utilized by being fed back into the raw material supply system. But first the material must be ground so that its size is similar to pellets, the common feed.

Another source of recycled material is material processed from articles after consumer use and generally referred to as postconsumer recycled materials or PCR. Recycling laws, applied mostly to packaging material, and practices in many communities require collecting and reusing such materials and sometime mandate the use of PCR in new products. The management and handling of PCR includes washing and grinding it before it is returned to a processor. In applications involving food packaging, a more rigorous cleaning system is required before the material can be reused. The processes

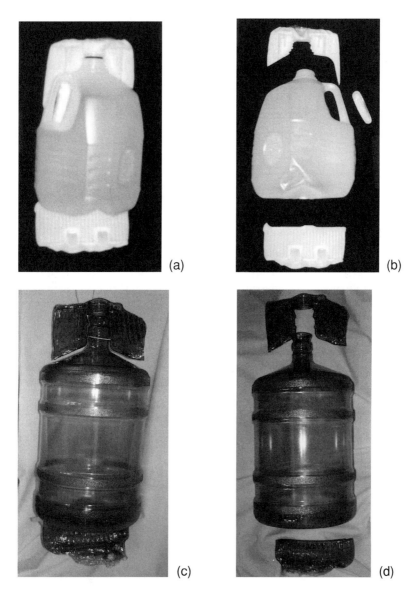

Figure 7.1 One-liter milk bottle (a) as formed, including pinched areas, and (b) after trim separation. Five-gallon water container (c) with pinched area and (d) after trim separation.

(a)

(b)

Figure 7.2 Blow molded automotive air duct: (a) as molded, with pinched off areas, and (b) final part. (Courtesy of Krupp Plastics & Rubber Machinery, Inc.)

used for such cleaning and the cleaned up materials must be tested to determine the success in removing various contaminants. Such data is presented to the proper authorities for approval. Only material from an approved recycler should be used to produce food packages. An alternate method is to limit the use of PCR to an inside layer in the wall by using coextrusion.

The common denominator for all those streams of materials is that the regrind produced is fed back into the extruder after it is mixed with virgin pellets to produce new blow molded parts. It is likely that other additives, such as colorants or stabilizers, are also added to the virgin regrind mix at the same time that it is prepared for extrusion.

7.1   REGRIND AND PCR USE CONSIDERATIONS

Two major areas, performance and contamination, must be considered before using regrind in an application regardless of the economic impact that such use has on a particular job. Once the decision is made in favor of using regrind or PCR the plant material handling systems must to be engineered properly to accomplish the task.

7.1.1 Performance

Most processes for producing plastic articles, and especially those which first melt the material, involve heat and shear. Such exposure results in some degradation of material properties or performance characteristics. The degradation results from the breakup of the molecular structure creating shorter molecules. Properties of plastic materials are directly related to the molecular weight or the molecules' length. Improper processing like excessive heating of the plastics, very high shear, or long residence time in an extruder at high temperatures, results in high levels of degradation and decline in performance. Typical of such changes are reduced strength of the material as measured by ultimate tensile strength and reduced elongation that the material sustains before breaking. Both together result in reduced impact resistance.

Expected part performance must be predicated on the regrind level and source, rather than just on virgin material capabilities. A common mistake during development is the use of virgin material only, although regrind is to be used in production. Additionally, development work that is done on a scale smaller than the production equipment is most likely to be less abusive to the material being processed. Virgin material and small-scale machines combined may lead to performance levels which are not achievable in full-scale production. In such cases a full-scale evaluation is in order, to establish the performance capabilities of the commercial scale production parts.

If the planned mixture fails to provide the desired performance, consideration must be given to limiting the regrind to a lower ratio even if the regrind generated through normal operation exceeds the system's ability to consume it. Higher performance virgin material could possibly provide an acceptable margin for adding more regrind. Very demanding applications may allow no regrind use.

The material mix used for determining product performance must first reach a steady-state composition. Every additional pass of the virgin/regrind mix through the extruder creates small fractions of material that have a successively higher number of passes through the process, hence further reduction in performance, as illustrated in Table 7.1.

For this example it is assumed that the process starts by feeding virgin material into the extruder. As regrind becomes available, it is mixed at a rate of 25% of the total feed.

The general equation to calculate the mix fractions at steady-state operation is

$$f_n = VF \times (RR)^n$$

where f_n = nth pass regrind content in the mix.
n = pass number (number of passes through the extruder for the fraction)
VF = Virgin fraction
RR = Regrind ratio [= $(1 - VF)$]

All are expressed as a fraction between 0 and 1 (i.e., 0.1 = 10%).

Table 7.1
Passes Through Extruder

| Component | Virgin | Regrind content | | | |
		First pass	Second pass	Third pass	Fourth pass
Pass 1	100%				
Pass 2	75%	25%			
Pass 3	75%	18.75%	6.25%		
Pass 4	75%	18.75%	4.69%	1.56%	
Pass 5	75%	18.75%	4.69%	1.17%	0.39%

The equation and the example illustrate that the fractions higher than the fourth pass account for less than 1% of the feed, and therefore evaluating material containing up to three- or four-pass regrind is sufficient. For a particular case the applicable mix fraction can be determined and put into the equation to reconstruct a similar table. Where the regrind ratio is low, its effect is small. A large ratio of regrind will have more passes with a significant level of regrind in the steady-state mix.

7.1.2 Contamination

Regardless of its source, contamination disrupts the operation and can cause significant time, material and production opportunity losses. Undetected during production and part inspection, it could cause product failure and additional liability to the manufacturer. Although on occasions contamination can result from the virgin material, in many more cases its origin is the regrind.

Handling all raw materials and especially regrind requires good housekeeping practices with several checks and balances. No gaylord of material should be placed into use without an operator or a material handler first verifying that the ticket identifies the box content as the material called for and then looking into the box for a visual check that the contents agree with the ticket description. Such a check easily confirms that the color is right, and allows noticing obvious contamination. It is very easy for dirt, garbage, and other foreign materials to end up inside a gaylord especially if it was left uncovered while the material was in storage or during previous use. Additionally, the operator should sift through the material manually and watch for any matter that looks different from the pellets or the regrind expected to be there.

Once contaminated material is mixed up with virgin material the total unusable material is significantly more than just the regrind part, and there is no chance for separation or recovery. This leads to the first system design solution: keep the streams of materials separated as far down stream as possible and preferably into the throat of the extruder. When a system is built in such a manner, any stream of material can be stopped immediately without taking the entire line down or with only small amount of material that needs to be discarded.

A very potent source of contamination of regrind within the plant is the mixing of different materials. This problem may be more acute where frequent changeovers occur on the same production line. Lingering fines and dust from one material in raw material transfer lines, hoppers, or loaders, are sufficient to cause major contamination of the subsequent run.

The use of various materials or colors on the same production line requires a well-established procedure for cleaning all systems at each changeover. Such procedures may call for partial disassembly of some system components, for example, the resin conveying system or loaders, change of filters, and washing down hoppers and grinders. In extreme cases the easiest way to get all residue from the extruder is to pull the screw out and clean the barrel and the screw separately.

Another source of troublesome contamination is tramp metal. The source of the metal is any machinery used in the production, when metal pieces break off during operation. Occasionally screws and nuts are inadvertently left behind during maintenance. Regardless, such metal reaching the extruder can cause considerable damage.

Metal trapping or detection devices incorporated into the material flow stream just before the extruder provide a measure of protection to the system. The simplest device is a magnet positioned in the material flow stream at the bottom of the hopper (Fig. 7.3) or at the throat adapter (Fig. 7.4) between the extruder and the bottom of the hopper. Standard units are available from many suppliers for both arrangements. The feed throat magnet is the more desirable configuration. By providing a shut off at the bottom of the hopper above the magnet drawer, the flow of material can be stopped and the

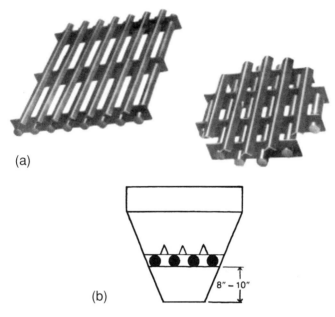

(a)

(b)

8″ – 10″

Figure 7.3 Hopper magnets: (a) rectangle and round designs; (b) most efficient placement of magnet in hopper. (Courtesy of Walton/ Stout, Inc.)

magnet pulled out for cleaning. During operation more and more particles are collected on the magnet. If they are not removed periodically, the material flowing by will start to pull them off the magnet back into the material stream and into the extruder. It is more difficult to reach the magnet at the bottom of the hopper and the hopper must be empty for cleaning to take place. Therefore, hopper magnets are more suited to short jobs where the magnets are pulled out and cleaned at each setup.

Metal detectors provide protection where the magnets are no longer effective, as they detect all metals, including nonmagnetic and stainless steel. Similar to magnets, they are installed in the flow stream preferably just before the material flows into the extruder. Metal is detected by applying an

Figure 7.4   Drawer magnet. (Courtesy of Direct-Line Products.)

electrical field which is affected by the presence of metal. When metal is detected, the flow is diverted for a short time, dumping the contaminated material into a collecting bin (Fig. 7.5). The small loss of material along with the metal is a low price to pay to protect the extruder, the extrusion head, and the blowing system, including the molds.

7.2   MATERIAL HANDLING SYSTEMS

It is clear that the use of regrind or PCR in any operation requires that the material handling system be designed with this goal in mind. Plant personnel must be well trained and versed in the proper procedures to handle multiple materials and multifeed streams into the same machine.

The preparation of recycled materials for return into production always includes grinding. In some cases the preparation of the material can include repelletizing after grinding. This is advisable where the regrind level is high because it

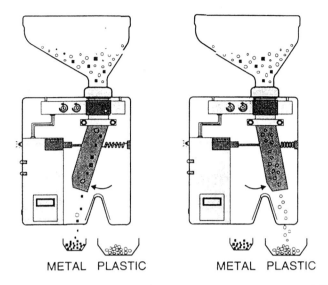

Figure 7.5 Metal detector and separator. (Courtesy of T+T Technology.)

assures better particle uniformity and stable feed and extruder operation. PCR usually undergoes washing and drying first. The washing commences after grinding. The regrind needs to be dried before it is stored for use or before it is pelletized.

Although repelletizing is preferable from the viewpoint of mix and feed it adds cost and an additional heat history with associated degradation. The degradation during pelletizing is normally less that the level experienced while going through the process because of the specialized design of the extruders used for this purpose.

7.3 GRINDING

The quality of the regrind depends on the operation of the grinder and the material being ground. Although there is no easily identifiable measure to determine the quality of the re-grind, high quality refers to uniformity of particle size, the least amount of dust, stringers, and other unusual particles, freedom from any contamination, and minimal drop in properties.

Grinding difficulty differs from one material to another and depends on material toughness and the size of the pieces fed into the grinder. Other parameters that play a major role are the design of the grinder, its mechanical shape, and especially the sharpness of the blades and the bed knifes.

Dull blades or bed knifes increase the variability of the particles after grinding but even more importantly generate large amounts of dust. Improper operation also generates stringers, long slivers of material.

Dust or very fine particles sometimes called just "fines," when fed into the extruder, do not melt properly or can even pass through the entire extruder without melting at all. Dust particles mixed with larger particles cannot generate the frictional heat required for melting in a extruder. The unmelted particles pass through the extruder and are in the melt that forms the part but have the same effect as contamination by foreign matter. Such inclusion can vary from just a visual defect to premature failures because they increase stress upon load or impact.

Stringers are likely to cause jam ups in the conveying lines when they hang up at various points in the system, especially at bends in the pipes.

Grinding the in-plant trim and reject can be handled by a dedicated grinder feeding directly back to the extruder or a centralized grinding system remote from the production machine. A dedicated grinder provides far less chance of contamination, less handling of the trim and rejects, and makes it easier to return the ground material into the same job. PCR's handled as a separate operation and should be treated as any other raw material supplied to the job.

The basic performance of the grinder depends on the choice of a particular design. Designs differ in grinding chamber geometry, the number and arrangement of the blades and the bed knives (Fig. 7.6), which control the feed of the material into the cutting area and the cutting action, respectively, and the specific design of the blades (Fig. 7.7). The final particle size coming out of the grinder is controlled by the screen size through which the regrind leaves the cutting chamber and the operating conditions of the grinder (1). The choice of grinder for an application must take into account first the material to be ground, the throughput required, the size of the pieces that will be fed in, and the required particle size of the regrind. Manufacturers of grinders can provide more detailed information along with the technical advice needed for proper selection.

The specific performance of a grinder depends mostly on its operating condition and its maintenance. A common operational problem is exceeding the grinder's rated capacity which results in overloading the motor and failure to obtain the designed rotor speed and therefore the cutting efficiencies or jamming and stalling of the grinder. Belts burn out when the motor keeps turning while the rotor is stuck. Overloading is caused by the material feed rate and also by pieces which are too large or too thick. Very large pieces, especially in industrial blow molding, may require precutting before pieces are fed into the grinder. Specially designed machines are offered to accomplish such precutting tasks.

The sharpness of the blades and their physical integrity must be maintained along with the normal maintenance of the grinder, like lubrication and belt inspection. The blades which are made of highly hardened steel to maintain their sharpness, are very brittle and are frequently damaged even in normal operation. Such damage is more likely to occur because of a grinder jam or because of tramp metal passing through. Regular maintenance requires periodical removal of the blades and bed knives for resharpening or replacement.

Proper reinstallation of blades and the bed knives is of utmost importance for good grinder performance. The clearance between the blades and the bed knives controls the cut-

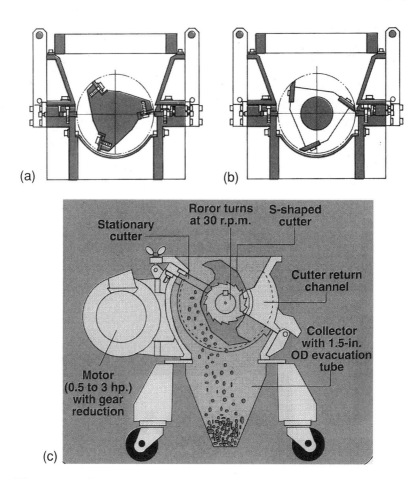

Figure 7.6 Basic designs of granulators. (a) Radial knife mounting on a solid rotor for heavy-duty jobs. (From Ref. 2.) (b) Tangential mounting with open rotor for lighter cutting. (From Ref. 2.) (c) Scutter granulator nibbles large pieces before final grinding. (From Ref. 3.) (d) Rapid granulator design for high capacity and handling a wide range of plastic parts.

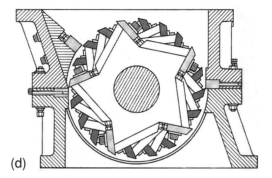

(d)

Figure 7.6   Continued

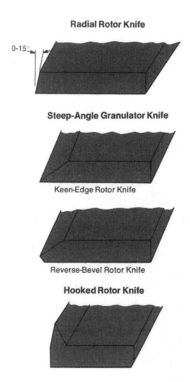

Figure 7.7   Common granulator rotor designs. (From Ref. 2.)

ting action of the grinder and its mechanical functionality and must be set precisely to the manufacturer's specifications. Too small a clearance results in jamming the grinder or even in metal to metal contact and extensive blade damage. Too large a gap results in inability to cut the material. Specially selected bolts are used in mounting the blades to the rotor and the bed knives to the grinder chamber to assure their survival under tough operating conditions and should not be replaced except with bolts with the same rating. Any locking arrangements for bolts to secure them from loosening or falling out during operation should be used according to the grinder manufacturer's instructions. Plant mechanics must be properly trained in setting the blades in the grinder and must closely follow the procedures established in the plant and adhere to all of the manufacturer's instructions.

7.4   FEED SYSTEM

A raw material feed system can be a very simple one of a hopper on top of the extruder which is fed manually or by a self-contained loader. Or the hopper could be charged by a loader which is a part of a plantwide material conveying system with centralized control and supply points.

Adding regrind and possibly colorant makes the line needs more complex. In manual systems the various components of the mix, the virgin material, the regrind, and the additives can be mixed manually off line in a drum tumbler or blender. A proportioning loader (Fig. 7.8), as an individual unit on the extruder, loads materials at a preset ratio. Some mixing action after loading in a mixing chamber or in the hopper, improves homogenization of the feed before the transfer into the extruder.

The preferred system is one that keeps the stream of virgin material separated from the regrind as far downstream as possible, meaning a small mix chamber and hopper at the

Figure 7.8   Hopper proportioning valve (Whitlock). (From Ref. 5.)

throat of the extruder with its control integrated with the extruder (Fig. 7.9). To minimize contamination, grinding trim
and rejects in a dedicated grinder next to the molding machine is preferred. Automatic transfer of the material from
the grinder into a holding hopper or box and feedback on demand into the extruder are also preferred. Monitoring the
flow of all components prevents the system from continuing to
operate when one of the components stops feeding and the
feed becomes all virgin, or worse all regrind, without anybody
noticing.

   Loading the mix ratio of virgin and regrind into a hopper
does not assure that the same mix is fed into the extruder, especially when a large hopper is used for inventorying material or when a drying hopper typically designed to hold material equivalent to four to six hours of extruder throughput is
used. A difference in particle size and shape between the virgin material and the regrind can cause separation between
the two components from normal flow in the hopper. The
smaller particles flow through the larger ones and pass
through the hopper at a higher rate and cause fluctuation in
the regrind to virgin ratio. This is more acute in a batch operation when the hopper is filled once before processing starts
and a large drop in inventory occurs between refilling.

   The change in the mix ratio has several effects on the extruder and the final product. The stability of melting in the

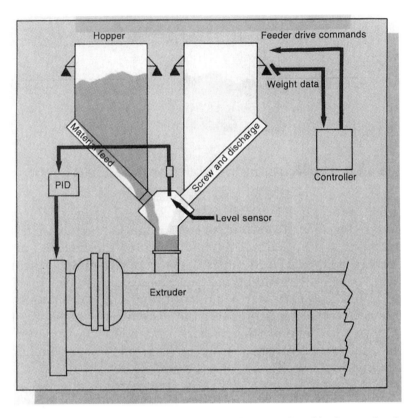

Figure 7.9 In gravimetric control, a level sensor is critical to maintaining throughput. (From Ref. 3.)

extruder depends on the uniformity of the solid feed (the pellets and regrind before melting) especially on the feed density and particle size which control the rate of melting. Variability of the feed and melting results in pressure fluctuations that translate through the extruder and cause pressure fluctuation at the extruder die and hence fluctuation in output. When fluctuations become pronounced they are called extruder surge. This variation in output is also associated with melt temperature fluctuations in the parison, which in turn affect the blowing.

Mix fluctuation also causes variation in part quality. First there is the effect of the regrind contribution to the physical properties described earlier. This effect can be very acute if the mix becomes occasionally rich with regrind. Secondly, product weight will vary with the fluctuation in output. The range of weight fluctuations must be limited so that low weight parts still meet performance and contractual obligations. Performance properties, such as impact and color, as discussed in the performance section, are the most sensitive to fluctuations.

For best results the system should minimize segregation of material after mixing by proper hopper design and by matching the particle size and shape of the regrind to the virgin material, as closely as possible. A tall slender hopper with a large L/D ratio (L is the hopper's length or height and D is its diameter), improves the flow characteristics in the hopper. However metered feed and mixing just before to entering the extruder is the best choice.

Many companies offer blenders which meter the components of the feed into the hopper or directly into the machine. The common design element for these systems is an individual feeder for each component which measures the throughput and a common mixing chamber to homogenize the mix before the processing machine.

There are two type of metering systems, volumetric and gravimetric. The volumetric system dispenses a fixed volume of each material. The volume unit of the blender is provided by a feeder, such as an auger or vibratory tray. The feeders are activated by a level control in the blending hopper and operate at the preset rate as long as there is demand for feed. Because volume is measured rather than actual weight, it is highly likely that the amount measured may vary because it depends on uniform bulk density of the components. Bulk density, given in lb per cubic feet, should not be confused with solid density. Balk density refers to the packing of pellets in a unit volume and can be found in material data sheets provided by suppliers. The value usually reported is the bulk density of the virgin pellets. The bulk density value for the regrind need to be established independently in each operation because it varies from plant to plant and possibly from line to line depending on grinding operations. In practice it means that the equipment setup may differ from line to line even when the same mix ratio is needed. The best approach is to calibrate the mixer based on the actual feed by measuring the output of each component and adjusting the relative settings accordingly. During operation the calibration should be rechecked frequently to confirm that the mix ratio is maintained.

Gravimetric blending systems circumvent variation in the bulk density of the components by weighing the components directly and adjusting the amounts via a feedback controller to the prescribed individual rate or the mix ratio. Because of direct measurement by weight rather than volume, the gravimetric system is more accurate and requires less setting up and maintenance during operation. The system is less susceptible to changes in feed characteristics, such as particle size, shape, or bulk density.

Two designs are common in gravimetric blenders, batch (gain-in-weight) and continuous (loss-in-weight) (Fig. 7.10). The loss-in-weight blender uses a feed system similar to the volumetric blender of an auger or vibratory tray except that the amount of material transferred from the hopper to the blend hopper is determined by load cells. Each ingredient hopper has its own weighing system connected to the unit controller.

The gain-in-weight system uses a batch hopper into which components of the mix are added sequentially. The amount of each component is determined by weighing the batch hopper and operating the discharge form the ingredient hopper. When all components of the mix are added, the batch is dropped into the mixing hopper for homogenization.

The operational sequence of the gain-in-weight blender limits its flow rate, and makes it more suited to low to moderate throughput, whereas the loss-in-weight (continuous) blender delivers high throughput.

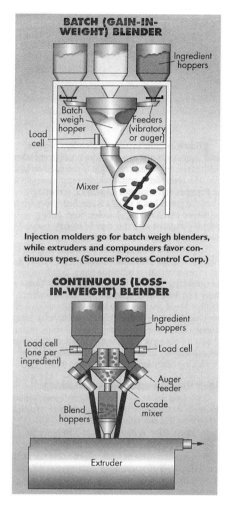

Figure 7.10 Examples of gravimetric blender design. (From Ref. 6.)

The electronic control system of gravimetric blenders offers an easy way to maintain the desired mix ratio for the entire production. Adjustments are as easy as adjusting the set points of the controller. However, the overall system is much more complex than a volumetric system or a proportioning loader.

7.5   EXTRUSION

Acceptable melt quality and extrusion of a uniform parison cycle after cycle are a must for a productive blow molding machine and line operation. Although the ultimate definition of melt quality may change from one operation to another, all include a well-mixed and highly uniform melt, especially in temperature, flow rate out of the die, color, and appearance within the part and among parts throughout a production run.

When regrind is used in the feed, the optimum conditions of the extrusion process are not likely to be the same as the optimum conditions for virgin resin alone. Optimization of extrusion requires a specifically designed screw, and therefore can be justified only where continuous operation at a specific feed mix is planned and maximum system performance is sought. In special cases, achieving the melt quality required for blowing the article, requires a dedicated screw.

General purpose machines may require a complement of different screws to accommodate different resins. A general purpose screw, which provides an acceptable performance in many cases, may limit the extruder's ability to melt a particular material or reach the needed melt quality. Of special concern are heat sensitive materials, such as PVC, which require special screws and other features on the extruder. Polypropylene, which is typically difficult to feed into the screw, may behave completely differently when mixed with regrind.

Many years of experience collected by extruder manufacturers and by companies specializing in screw design provides improved performance through revised screw designs. The use of any one of the available extrusion simulation computer programs and proper experimentation can also yield a optimized screw design for a particular material or feed. Such developments provides the proper screw and also helps to establish the proper operating parameters for the extruder, including zone temperatures and die head design.

REFERENCES

1. TJ Stuff, J Strebel. How grinder variables affect bulk density and flow properties of polyethylene powder. Plastic Engineering, August 1997.
2. MH Naitove. Buying granulators: No more a casual matter. Plastic Technology, May 1975.
3. DL Dunnington. Gravimetric extrusion control neutralizes effects of process variables. Modern Plastics, January 1992.
4. Advanced Rapid Plastics Technology, July 1986, p. 41.
5. SM Lemay. Resin-handling systems deliver a range of process and cost benefits. Modern Plastics, June 1993.
6. D Galante Block What to look for in a gravimetric blender. PT Auxiliaries, March 1997.
   V Wigotsky. Auxiliary equipment, Plastic Engineering, Feb 1997.

8

DEFLASHING, TRIMMING, AND REAMING

Alex Orlowsky and Anatoly Orlowsky

Seajay Manufacturing Corporation
Neptune, New Jersey

From its early beginnings the extrusion blow molding process has required that post operations on the blown plastic product.

Before handle ware, it was necessary to trim the tail and the moil from the blown plastic container (see Figs. 8.1 and 8.2).

In early blown plastic container production it was also necessary to ream the inner diameter (i.d.) of the neck finish to ensure that the inside diameter of the finish was round (not oval) and allowed for a plug style closure to be used.

Reaming of the finish caused angel hair and chips to dislodge and fall inside the container, necessitating that a vacuum be applied to remove them.

Figure 8.1   Impact deflashing. (Courtesy of Johnson Controls, Inc., Plastics Machinery Division, Manchester, MI.)

Figure 8.2   Off-line deflashing unit. (Courtesy of Seajay Manufacturing Corp., Neptune, NJ.)

The term flash is the excess plastic material along the parting line of a blown product. Flash is generally formed on the sides and indentations of a blown plastic product.

Initially, the operator removed this flash by manually tearing or cutting this excess material. Even though removing flash by this method is archaic, it is still done today in many industrial blow molding plants. Neither the custom blow molder or his ultimate customer wants to or can afford to invest in semiautomatic or automatic flash removal.

Initial blow molded plastic containers were round. Threaded openings were formed during the blow molding process. Initial problems of the blow molding process centered on achieving the desired dimensions to allow the closure to seal properly either on a smooth perpendicular surface or as a plug seal.

Therefore, a fixture was designed utilizing a straight metal blade in a guillotine to separate the moil from the required height of the blown plastic finish. The neck of the blown plastic container was placed through a hole and pressed manually until the shoulder of the bottle stopped the plastic container from entering any further and then the operator either used a foot pedal or a hand switch to activate the guillotine. This process was laborious, monotonous, and unsafe.

The method was effective, yet leaks occurred due to the imperfect alignment of the blown container within the fixture.

The guillotine method of detaching the moil required a second operation. This was fly cutting or reaming the inside diameter of the blown plastic finish. The guillotined blown plastic container was positioned under a stand and drill press and a hole cutter or drill was used to achieve the required inside diameter. This method was used to produce the early gallon milk HDPE containers. In some custom blow molding plants, holes are still made by using drill presses and drills or hole cutters.

The immediate success of plastic blow molded bottles led to almost daily improvements to in deflashing and neck finish trimming. Companies, such as Owens-Illinois, Continental Can, Dow Chemical, and Union Carbide, quickly began to automate the neck finish trimming and reaming and tail removal.

Improvements were constant and handle ware soon appeared in the market place for half gallon and gallon blow molded HDPE bleach containers. They were an immediate success. The handle ware presented the blow molder with two additional concerns. The collapsed or pinched material trapped to allow forming the blown handle and the excess flash due to the excess parison formed for the blown plastic handle both had to be removed. The early blown plastic handle ware was trimmed by the operator with a trim knife. Pinch-off designs were learned for the various plastics, such as HDPE, PVC, LDPE, PS, HIPS, and PP, and sometimes the flash and pinched plastic could be easily removed manually. However, once again, this was time-consuming, laborious, and monotonous.

With the advent of blown plastic handle ware, four steps were required to ready a commercial plastic container for the user.

Blow molders were faced with new resins, relatively untried machines, short runs, and untrained work forces. The blow molding industry was growing at a 10–15% yearly rate.

There was an immediate need to produce post-operated equipment that could remove the tail, excess flash, cut the moil, and remove the pinched handle material in one or two operations automatically, yet be flexible enough to handle different container sizes and shapes.

Seajay came into existence and built the first commercial integral trimmer/reamer. It was operated pneumatically and in a semiautomatic mode. A nest fixture that allowed the blown plastic container to be cradled and supported was used for each specific container design, and this method is still used today. The stroke of the guillotine and the nest fixture are adjustable and removable allowing for a trimmer/reamer to be cost effective (Fig. 8.3).

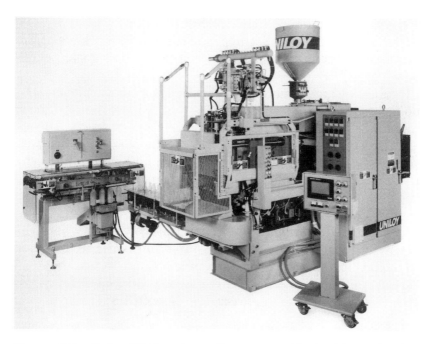

Figure 8.3 Unilox 2016 reciprocating screw with detabber/spin trimmer for 1-quart nonhandle container, four heads. (Courtesy of Johnson Controls, Inc., Plastics Machinery Division, Manchester, MI.)

Acceptance of blown plastic containers continued to grow. Production levels soared, blow molding machines began utilizing multiple blowing stations, and wheels became more dominant for large volume production requirements. Cycle times decreased with improved resins, better bottle designs, improved parison programmers, and new mold designs. This created the demand for in-line trimming. The term in-line trimming depicts the entire trimming process incorporated within the blow mold versus downstream conveyor equipment. The operator was no longer required to do manual or semiautomatic trimming but now inspected and packed the product.

In-line trimmers utilize many of the original devices, such as trim nests, guillotines, and reamers, and the container passes by each post operation via pneumatic gating cylinders.

At the same time new technologies were developed that do not require reaming of the finish, called a "calibrated neck finish."

During the blow molding process, the blow pin enters the parison and compresses the hot plastic material to form the threads, lugs or other design and also forms the inside of the neck finish. The blow pins are tapered for ease of entry and removal and can be also water-cooled or air-cooled. A striker plate is also designed on the top of the blow mold where the blow pin enters, and as the blow pin enters, it is so designed with the blow pin dimensions to pinch off the excess material (moil), for removal within the blow molding machine or for downstream guillotining to create a smooth, flat surface.

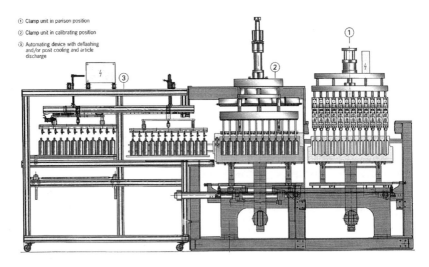

Figure 8.4 Clamp units in different positions. (Courtesy of K.P.R.M. (USA), Inc., W. Müller Division, North Branch, NJ.)

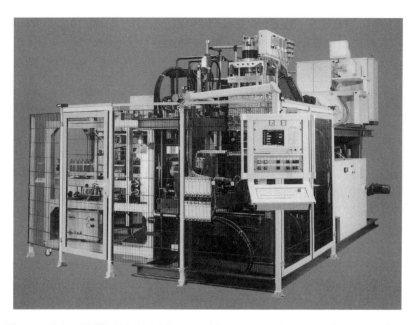

Figure 8.5 FMB 4/6-100 blow molding machine: extruder, 70/24L/D; head, s7/35-85 P-PE I Ch. W/C; article, 250-ml HDPE Unicara cosmetic bottles; net weight, 25 g; output, 2300 pcs/h. (Courtesy of K.P.R.M. (USA), Inc., W. Müller Division, North Branch, NJ.)

Spin trimming is also utilized whereby the blown moil is removed by rotating the blown bottle on counterrotating belts and a correctly positioned razor knife cuts off the moil like a can opener (see Fig. 8.3).

A variety of techniques are employed today in the blow molding industry for trimming, reaming, and deflashing. The need for flexible, well-fabricated, downstream trimming is essential. Attempts to produce trimming equipment located on the manufacturing floor have proved to be futile.

New trimming methods are constantly being designed, including laser trimming by using a programmable control.

Trimming of each product requires the proper pinch-off design, quality mold design and construction, correct speeds of knives, drills, and guillotines, all matched to the specific plastic material used in the blow molded product.

ACKNOWLEDGMENTS

Seajay Manufacturing Corporation, Neptune, NJ; Heins PCM Machinery Ltd., Brampton, Ontario, Canada; Proco Machinery, Inc., Mississauga, Ontario, Canada; Uniloy, Johnson Controls, Inc., Manchester, MI; Bekum America Corporation, Williamston, MI; K.P.R.M. (USA), Inc., W. Müller Division, North Branch, NJ.

9

TESTING OF EXTRUSION BLOW MOLDING

Cheryl L. Hayek

TopWave International, Inc.
Marietta, Georgia

9.1 INTRODUCTION

9.1.1 Faster, Cheaper, Lighter, Stronger: Good Quality

Plastics have gained popularity due to advantages, such as safety, convenience, lightweight, ease of handling and transportation, that they afford the end product. Everyday there is more demand placed on manufacturers to exceed current levels of performance, quality and turnaround time while reducing costs. Plastic processors are being asked to push the limits in applications in the medical, packaging, automotive, appliance and other industries. This has led to the modification of existing polymers and development of new plastics. It is important for designers, engineers, and manufacturers to know the inherent characteristics of these materials when

219

considering the appropriate one for a specific product. In addition to testing the polymers to establish standards for evaluating material selection for a given application, a battery of end-product testing must also take place to ensure product integrity throughout the life of the part or product. The specific tests depend on the plastic product and the parts' intended use. This chapter overviews some of the most common tests used for extrusion blow molded parts.

Testing must begin with incoming material inspection followed by in-process and finished product testing. Incoming material inspection prevents and reduces wasted material and time encountered with out-of-spec materials. It ensures consistent quality of materials to be processed and eliminates downtime and maintenance that arise as a result of poor quality resin.

9.2   POLYMER TESTING

It is important to understand the standard tests used to characterize materials and how to utilize this data along with process and design knowledge.

ASTM, UL, DOT, and other standards do not necessarily correlate with end-product performance. Fabricating conditions and part design strongly influence the actual performance of a plastic part. Test methods have been a standard in the plastics industry as a basic measure of a material's physical, mechanical, thermal and process-related properties. Standards provide a good base to determine the strength, toughness, and processability of a material, which should be considered during material selection. The results of these tests are subjective and provide minimal correlation to the final part's performance.

9.2.1   Melt Index Test (ASTM D-1238)

This test measures the relative flowability of the resin when melted. It is used to determine the quality of the resin and is a postprocessing check. The flow rate is measured by an extrusion plastometer. The material is heated and extruded through a die at a designated length and diameter at a specific temperature, load, and position.

9.2.2 Capillary Rheometer Test (ASTM D-3835)

This test measures the flowability of the resin under a wide range of shear stresses and shear rates to determine the shear sensitivity and flow characteristics at processing shear rates. The same extrusion plastometer used for melt index is needed with an added mechanically driven piston. Values derived from the test are plotted on a flow curve.

9.2.3 Tensile Properties (ASTM D-638)

Tensile strength is a measure of the resistance of the material to stresses pulling in opposite directions and the extent to which the material stretches before failure. It determines the strength of a material by measuring the energy required to pull the test specimen apart and records the amount of stretch which has occurred. A bar specimen of the material to be tested is prepared. Both ends of the specimen are positioned in the jaws of the test apparatus and then pulled at rates of 0.020 in./min to 20 in./min. The tensile strength is calculated from the following formula:

$$\text{Tensile strength (psi)} = \frac{\text{Load (lb)}}{\text{Cross-sectional area (sq. in.)}}$$

9.2.4 Compression Properties (ASTM D-695)

Compression is the ability of a material to withstand compressive forces. A compression test helps to determine the overall strength of the material. It measures the stress required to rupture or deform the sample at a certain percentage of its height. A compression tool applies a load at a constant rate of movement to the test specimen. Compressive characteristics include elasticity, yield stress, deformation beyond yield, compressive strength, and compressive strain. Compressive strength is calculated as follows:

$$\text{Compressive strength (psi)} = \frac{\text{Load (lb) to failure}}{\text{Width} \times \text{thickness (sq. in.)}}$$

9.2.5 Flexural Properties (ASTM D-790)

This test measures the stress required to stretch the outer surface of the test specimen.

9.2.6 Izod Impact Testing (ASTM D-256)

Izod Impact is the ability of a material specimen to withstand a sharp blow. The specimen is prepared (18 in. × 1/2 in. × 2-1/2 in.) A notch is cut, and the sample is conditioned. The sample is inserted in the test apparatus with the notch facing the pendulum's impact, and the reading is recorded. The energy consumed in breaking the specimen is calculated from the height the pendulum reaches upon its return. Izod is an indication of overall toughness and flexural impact strength. The Charpy Impact test, similar except for positioning of the test specimen, is a good test for determining the best material to use when molding sharp corners.

9.2.7 Tensile Impact (ASTM D-1822)

This is a high-speed impact test. The test specimen is exposed to sufficient energy to cause failure.

9.2.8 Creep (ASTM 2990)

This test measures the changes a material exhibits while subject to a certain load at a certain temperature for a certain amount of time. It is measured in terms of tensile creep, compressive creep, and flexural creep.

9.2.9   Indentation (Durometer) Hardness (ASTM 2240)

This test measures the indentation hardness of soft materials. Readings should be taken on smooth, even surfaces, and the tests should be run at ambient temperatures. The sample is inserted, firm contact is made, and dial readings are taken one second after firm contact. The Barcol Tester is also used for determing the hardness of rigid plastics. It is a hand-held impressor with a spring-loaded plunger. Both tests measure the depth of indentation. Indentation hardness testing provides a quick understanding of possible mechanical property changes which can be affected by material processing and aging.

9.2.10   Deflection Temperature (ASTM D-648)

This test provides a gauge of the temperature at which minimal deflection occurs.

9.2.11   Torsion Stiffness Test (ASTM D-1043)

This test measures the relative change in stiffness of plastics over a range of temperatures via the angular deflections resulting from torque imposed at a specific range.

Physical properties may vary from the results of material testing and specifications provided by the resin suppliers because of factors, such as time of melt in the mold, lot to lot variation in resin, knit lines, amount of regrind added, and process conditions. In addition, end products are subject to various temperatures, stresses, and environmental conditions during their life that significantly influence part performance. This makes it imperative to do testing on the final part.

9.3   CONTAINERS/PACKAGING

Food and beverage, toiletries, cosmetics, health care products, medical, household chemicals and many others products make plastic packaging the largest segment of worldwide plastic consumption. Everyday more products are being converted to plastic from glass, aluminum, tin, paper and other materials. Lightweightness is one of the major advantages, in addition to ease of use, flexibility of design allowing for brand recognition, and reduced transportation, and handling costs. Testing is imperative to ensure product integrity throughout the product life cycle. The container is a vessel and a marketing tool that can lead to the success or failure of a product (see Table 9.1). Measurements involve two types of data, attribute and variable (see Table 9.2). Attribute data include measurements of data on random defects. These are defects that are mutually exclusive. They are caused by random occurrences during the manufacture of a product. Variable data are measurements that result from drift in process parameters, which if left unattended will continue to exist and possibly worsen.

Table 9.1

HDPE Container Characteristics

Critical measurements	Noncritical measurements[a]
Weight	Wall thickness
Top load	E dimension average
T dimension average	Flatness/saddle/planarity of closure
T average ovality	sealing surface
S and H dimension average	Overall bottle height
Z dimension land thickness	Bottle width
A diameter average	Bottle depth
Drop test	ESCR test
Capacity	Water marks
Seam leakers	Streaks
Delamination at parting line	Die lines
Excessive/incomplete mills	Color match
Pinhole leaks	Damaged threads

[a]Includes attribute data (color, visual defects, etc).

Table 9.2

Examples of Attribute and Variable Data

Attribute data	Variable data
Black specks	Body dimensions (width, depth,
Bubbles	taper, etc.)
Unmelted resin	Ovality
Leaks	Wall thickness
Strings	Overall height
Neck folds	Perpendicularity
Seal surface defects	Finish slant, cocked neck
Mechanical damage in neck area	Neck finish dimensions
Mechanical damage on sidewall	(T, E, H, Z, A, etc.)
Fish eyes	Top load
	Volume
	Base height, clearance

9.3.1   Finish Dimensions

Finish dimensions are critical to ease of handling during filling and capping operations. These measurements are also critical to safety and liability and to end-user perception, i.e., tamper evident seals, ease of use, and proper closure. (Fig. 9.1). Dial calipers, go/no-go gauges, pi-tape, optical comparators, and camera-based (CCD) dimensional systems are used for measurements.

9.3.2   Body Dimensions

Measure the critical dimensions, such as overall height, outside diameter, and shoulder contour, to ensure compliance to specifications and to ensure trouble-free handling during filling and capping operations. The same tools used for finish dimensions are used for these critical dimensions.

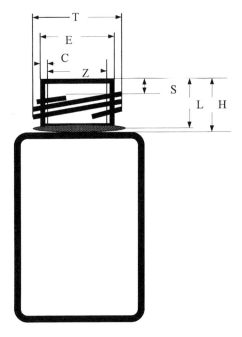

Figure 9.1   Neck finish dimensions.

9.3.3 Perpendicularity

This test ensures that the container stands upright relative to the horizontal surface on which it stands without any deviation from the vertical. The container resting on its base is rotated 360° against a centering device or height gauge, and measurements are taken over the maximum range (see Fig. 9.2). Perpendicularity measurement is important to avoid damage to filler vent tubes. In addition, excessive tilt or rocker bottoms contribute significantly to poor package appearance, which directly affects the consumer's perception of the quality of the product it contains.

9.3.4 Weight

A laboratory balance is used whose capacity is 100 grams minimum and whose accuracy is ±0.1 gram. The Operator

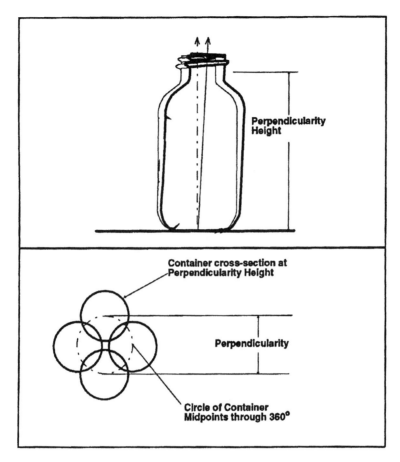

Figure 9.2 Perpendicularity measurement.

must inspect for and remove any foreign material before weighing the container. The average, range, and standard deviation are recorded for the sample group. Characteristics, such as water vapor transmission and strength, are influenced by container weight.

9.3.5 Capacity

Brimful and fill point capacities are measured from the top of
the finish down to determine the internal volume of a container. Fillers are required by law to guarantee that bottles
contain as least the volume stated on the label. Capacity is
measured by using a balance, medicine dropper, and ther-
mometer. Measurement is accomplished as follows: first
record the tare weight of the empty container, then fill it with
room temperature water at a designated fill height, using a
medicine dropper or fill point syringe until the convex meniscus is precisely at the correct fill height. The filled container is
weighed, and the temperature of the water noted, taking into
consideration the density compensation from a table of density versus temperature. Instruments are available now that
automate this traditionally manual test, eliminating the effect of operator variability and errors inherent in manual
recording.

9.3.6 Thickness Distribution

Successful container performance depends heavily on proper
wall thickness distribution. Critical properties, such as water
vapor transmission rate, top load strength, and volume
change, depend on wall thickness. Various levels of instru-
mentation available for measurement include micrometers
and calipers, section weight analysis using hot wire cutters,
ultrasonic or Hall effect gauges, optical comparators, and automated IR absorption measurement systems (Fig. 9.3). These
are the important areas of container wall thickness measurement to (Fig. 9.4):

> Base
> Base corner/heel—the lower major diameter
> Sidewall—body label area
> Shoulder-upper label panel area major diameter
> Upper shoulder—curvature above the label panel

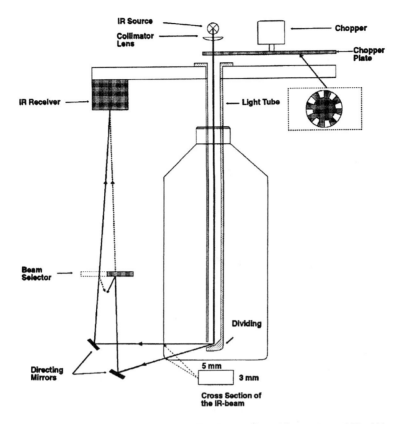

Figure 9.3 Wall thickness using IR absorption. (Courtesy of TopWave International, Marietta, GA.)

9.3.7 Sidewall Rigidity

The resistance to deformation of the sidewall under an externally induced pressure, i.e., hand grip, will be equal to or greater than a specified force. This resistance is defined as the minimum force necessary (when applied at the midpoint of the sidewall) to raise the liquid contents fill height from the fill point level to overflow by inward flexure.

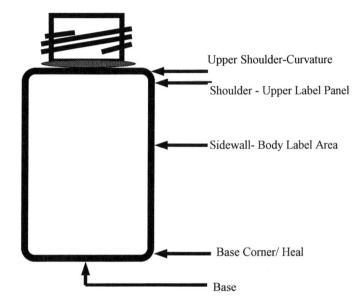

Figure 9.4   Wall thickness measurement areas.

9.3.8   Container Distortion

Several factors affect the shape of a filled container that can be anticipated and addressed during bottle design. Design techniques, such as ribbing, tapering, recessed panels, horizontal flutes, and diameter changes help overcome vacuum and partial vacuum caused by oxidation, ultraviolet light, hot fill, permeation, and swelling. Vacuum properties (the amount of vacuum necessary to cause paneling of the container wall) can be measured and are often done so using in-house vacuum testing devices.

9.3.9   Drop Impact

This test simulates the degree of abuse that a container must withstand under varying conditions (normal to extreme) to which it will be subjected during its life. The test is accomplished by filling a container with the actual product or contents having a similar specific gravity and dropping it from various heights at various angles onto a hard platform and measuring the maximum height of drop without failure. Samples are strictly conditioned at varying temperatures to approximate environmental factors to which the container will be exposed. Hold the container at a 30° angle at a designated height and release it so that the container free falls onto the concrete drop area. Visually check each container and record any deformations, such as cracks, rocker bottoms, and breakage. It is extremely important to indicate the exact position of each specific type of stress failure to find out precisely why a container has failed. Keep in mind that plastics are chosen over other packaging materials partly because of their ability to resist breakage, avoiding harm to the contents and to the container user. The drop test confirms proper material selection and container design. Results may indicate the need for redesign, better wall distribution, or increased container weight. Impact resistance is usually greater in container designs whose thickness transition is gradual and complex detail is minimal.

9.3.10 Top Load Testing

Top load testing is a measure of the vertical strength of a container. It is essential to ensure that the container withstands various loads imposed throughout the product's life. These include storage, transportation, stacking and load requirements, and top load impact during filling and capping. No container can crack, yield, or buckle under the specified top load.

9.3.11 Appearance

A visual check of the container is of great importance to its aesthetics and function. Control of appearance attributes is imperative to the success of the product at the point of sale. Inferior package appearance makes the consumer feel that the product is inferior. The operator/technician should check each sample for orange peel, parting lines, scratches, blemishes, paneling, deformation, color variation, and other surface irregularities, as well as inhomogeneity, such as holes, bubbles, fish eyes, voids, and contamination. As a quick check, a flat, black background with fluorescent lighting is used. Visual comparisons against standards are used for color, clarity, pearlescence, and haze. Visual checks are often done by operators whereas random checks, usually more comprehensive in nature, are done by QC personnel. Vision systems for 100% on-line testing are available for go/no-go inspection.

9.3.12   Surface Treatment

Polyethylene containers must be flame treated to ensure good label adhesion. Optimum flaming conditions are determined by using a stain test procedure, whereas routine quality checks are done by a water test. The flame-treated container is soaked in a solution containing diluted carbol fuchsin. The container is removed and when dry is evaluated for stain intensity and stain uniformity. The more pronounced the dye, the better the surface tension or flame treatment. The test sample is flame treated then placed under running water. If the water is evenly distributed over the surface area and holds for 20 seconds the treatment is good.

9.3.13   Environmental Stress Cracking (ASTM D-1693A, D-256B)

Environmental stress cracking (ESCR), a long-term measure of strength is measured in time to failure. Variables to consider when looking at ESCR are package design, material selection, contents to be packaged, temperature, fill level, and stress level. ESCR is measured by filling a sample container to various fill levels using a burette for accuracy. Then, the container is capped using consistent torque in the range of 10–20 ft/lb (13.6–27.2 joules). The containers are inverted, checked for leaks, and then stored in an environmental chamber at 140°F ± 1°F (60°C ± 6°C). The containers are visually inspected twice each day over a period of time. In research and development, the container is evaluated over longer periods to determine the point at which 50% of the samples fail. However, standard measurement is looked at as no failures within seven days.

9.3.14 Chemical Resistance (ASTM D-543, D-2552)

Chemical resistance is very important to the end customer. It is important to know the interaction of the package with its contents whether it is a food product, medicine, cleaning fluid, or hazardous chemical. The container performance is influenced by environmental conditions during its life, such as UV exposure, high to low temperature, moisture, and other weathering factors. Immersion tests followed by visual inspections and weight gain determination are often used in addition to ESCR and various tensile tests for chemical resistance.

9.3.15 Infrared (IR) Spectrascopy

Infrared spectrascopy measures the spectral absorption in the IR range using pyrolysis, transmission, and surface reflectance methods to identify types of plastics, surface coatings, and to determine wall thickness. This method is based upon interaction of infrared electromagnetic radiation with mass, resulting in absorption of certain wavelengths of radiation. This technique has been very popular, and new developments are expected to give rise to more applications for process testing.

9.4 LARGE PARTS AND INDUSTRIAL BLOW MOLDING

Because of the lightweightness and durability to withstand multiple trips, 55-gallon plastic drums have significantly replaced metal drums for hazardous materials and other products. Some of the most common methods used to test drums and other large extrusion blow molded products used in automotive, recreational, industrial and other applications are described below.

9.4.1 Drop Test

The operator fills a test container with glycol up to 98% of its outage capacity and conditions it at 0°F. A drop test is performed ranging from 4 ft to 12.8 ft. depending on the specific gravity that the drum is certified to hold, i.e., a 30–55 gallon drum, 1.9 specific gravity. Following the drop test, the container is inspected for any cracks, stresses, or deformation.

9.4.2 Leak Proof Test

This test is performed on drums containing liquids. An operator places an air fitting onto the side of container and inserts plugs and caps. Then, the test sample is submerged under water to 3–5 psi. The occurrence of any leaks indicates failure.

9.4.3 Hydrostatic Test

An operator selects five test samples, places a fitting into each sidewall, and fills them to the top with water. The test samples are plugged and placed on their sides, then pressurized with water for 30 minutes, according to the Department of Transportation Group 1, 2, and 3 or customer supplied specification. Any leakage is considered a failure.

9.4.4 Vibration Test

An operator selects three test samples, fills them with water, and places them on a vibration table for one hour. Then the test samples are placed on their sides immediately and checked for leaks. The container passes if there is no sign of leakage or rupture.

9.4.5 Stacking Test

This test determines how many drums or containers can be stacked in a three-meter area without compromising the integrity of the product An operator selects three test samples and fills each one with water up to 98% of its outage capacity. Then, the test samples are subject to a known load, 55 gallon drum to 2200 pounds, applied to the top head and stored for 28 days (for jerricans and drums intended to contain liquid) at 104°F. Nonliquid-bearing drums and containers are stored for a period of 24 hours. The second part of the stacking test takes place immediately following the first. An operator takes two like containers, fills them with water, and places them on top of a test sample. The containers must remain stacked for one hour without any sign of deformation or lost strength that would affect transportation safety or the products' integrity.

9.4.6 Static Leak

An operator fills a test sample to 98% of its outage capacity. Then, the test sample is placed on its side for a period of two hours. Any evidence of leakage renders the test a failure.

9.4.7 Thickness

Ultrasonic testing is used to measure wall thickness and to identify characteristics, such as delaminations, porosity, weld penetration, cracks, inclusions, voids, or joint integrity.

9.4.8 Ambient Drop Test

An operator fills a container with water, then torques it, caps it, plugs it, and submits it to a series of various drops. Visual checks are done for leakage resulting in failure.

9.4.9 Dynamic Compression

This tests the baseline of wall strength for drums/containers intended for liquid contents. A known load is applied to the container, and the corresponding deflection is recorded. This can also be done in the reverse by running the test to a certain deflection and recording the corresponding load.

9.4.10   Leak Proof Test: ESCR Stress Crack

An operator fills a test sample with 2% of a Hydrapol wetting agent. Then, the test sample is pressurized to 2 psi air pressure while a 2200 pound load is applied in a 122°F environment. The samples must remain leak free for 14 days. Any leakage within 14 days signifies a design or process defect.

9.4.11   Flammability

The establishment of suitable standardized testing procedures has been an objective of the plastics industry, various government agencies, and major standards organizations, such as Underwriter's Laboratories (UL), American Society for Testing and Materials (ASTM), and the Federal Test Method Standard (FTMS), to name a few.

UL 94 is one of the most recognized flammability tests for classifying plastic materials used in appliances and other applications. There are various flammability ratings assigned to materials depending on a test specimen's performance when subject to horizontal and vertical burning tests. Following is a list of some current flammability tests established by recognized standard organizations used for classifying plastic materials (Table 9.3).

The results of the tests shown in Table 9.3 can be classified into the following categories:

- Burning: samples that essentially are consumed by flames when subject to the test conditions
- Self-extinguishing: samples that continue to burn for a minimal period of time upon removing the test flame, but extinguish before reaching a prescribed mark or before complete consumption of the test sample
- Non burning: samples that when subject to the test conditions, do not inflame or extinguish so quickly after removing the test flame that self-extinguishing time is a meaningless measure

Table 9.3

Flammability Tests

FTMS	Vertical	Flame resistance for difficult to ignite plastics	Ignition time, burning time, flame travel
UL 94V-0	Vertical	Flammability of plastic materials	Rate of burning
UL 94-HB	Horizontal	Flammability of plastic materials	Rate of burning
ASTM D-568	Vertical	Rate of burning flexible plastics equal to or less than 0.050 in.	Burning rate, average time and extent of burning
ASTM D-635	Horizontal	Rate of burning rigid plastics over 0.050 in.	Burning rate, average time and extent of burning
ASTM D-757	90° to source	Incandescence resistance of rigid plastics	Burning rate, average time and extent of burning
ASTM D-1929	Horizontal	Ignition properties of plastics	Flash and self-ignition temperatures, visual
ASTM D-2863	Vertical	Oxygen index flammability test	Oxygen index
ASTM D-84	Horizontal	Surface burning characteristics of building material	Flame spread index, smoke density
ASTM D-3014	Vertical	Flame height, time of burning, and weight loss	Weight loss and time to extinguish

Source: Adapted from Ref. 15.

It is important to understand fully the test conditions pre-scribed by the standard methods presented previously. You are encouraged to contact the respective standards organiza-tion for specific details and to inquire as to the recommended tests to evaluate finished goods under representative or actu-al performance conditions.

9.4.12 Sound Tests, Noise, and Vibration

The part is fixtured into a vibration environmental chamber and exposed to varying frequencies and amplitudes at vary-ing temperatures. Then, the part is removed and visually checked for defects, such as, cracks, holes, weld breaks, and in-serts. This test is used on various assemblies, interior and ex-terior reservoir tanks, and climate controls.

9.4.13 Condensation

Condensation on climate control ducting for heating and air conditioning units is a problem for automobile manufactur-ers. Parts are placed in a humidity chamber and monitored for drips.

9.4.14 Stereolithography

Rapid prototyping of parts before final designs are made for production enables evaluation and preproduction testing of part performance, to reduce the potential for liability which is particularly important with high-value products. Stereolitho-graphy offers an improved method of creating product proto-types for early customer testing and for making temporary and production tooling. Turnaround time on the part and quality is greatly improved. With the design data coming from CAD and the test part made by stereolithography, the chances of quality problems in the dimensional integrity of the final product are greatly reduced. Using stereolithography for ear-ly testing also saves time and money by solving design issues early in the production cycle.

9.5 THICKNESS TESTING

Although thickness measurements taken by calipers and micrometers are standard forms of thickness measurement throughout the industry, they do not provide the most accurate or reliable results. Sample preparation is tedious, poses a threat of injury, and the possibility of faulty data. Measurements are subject to operator variation due to measurement technique (including pressure application and angular positioning of the instrument), interpretation of thickness readings, and transpositional errors encountered during measurement recording. These factors are eliminated with the nondestructive measurement methods described following.

9.5.1 Ultrasonic Gauging

This test method uses high-frequency sound and the reflected sound energy (pulse echo) to determine wall thickness. A transducer is placed on the part and acoustically coupled using a transfer medium (usually water, glycerin, or propylene glycol) (see Fig. 9.5). A measurement is recorded based upon the time it takes for the ultrasonic sound energy to move through the test part. Ultrasonic gauges must be calibrated to the type of resin being measured. Ultrasonic gauges interact only with one side of the test sample, and they are very easy to use. Measurement is affected by shifts in temperatures (greater than 10°F), so operators should calibrate and measure at ambient or constant temperature. This method is typically selected for large, rigid, wall parts and designs not allowing access to both sides of the test material.

9.5.2 Hall Effect Gauging

This method of measurement is based on a phenomenon called the Hall effect that uses a Hall sensor and static magnetic field. A steel ball of a known mass is placed inside the plastic container or part, and the Hall probe is on the other side (see Fig. 9.6). The steel ball follows the location of the Hall probe as it is scanned over the test part and a compari-

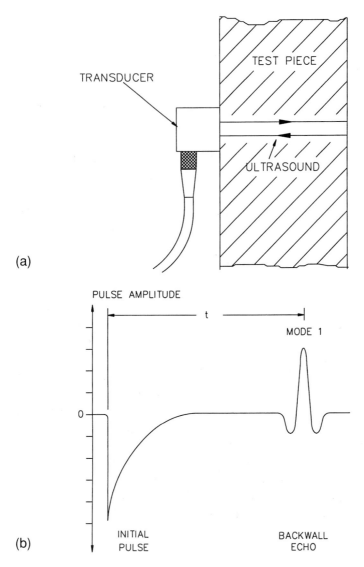

(a)

(b)

Figure 9.5 Wall thickness using ultrasound measurements: (a) test method illustration; (b) signal illustration.

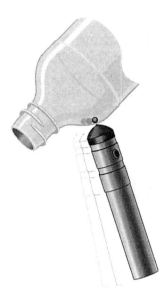

Figure 9.6   Wall thickness using hall effect measurement.

son between the induced voltage and the distance of the steel
ball from the probe results in a thickness reading. Hall effect
gauging is quick, easy, and especially suited to complex
shapes that have tight radii. It does not have to be calibrated
to the resin type being measured and does not use a coupling
agent. The angle of the probe is critical for accurate, repeatable measurement. This can be overcome through fixturing
devices and proper user training. This method is used
throughout the plastics industry for quickly measuring containers and other thin-walled parts used in medical, automotive, aeronautical, and recreational applications.

9.5.3   In-Mold Thickness Measurement

Maintenance of minimum wall thickness is critical for manufacturers of chemical storage drums, automotive fuel tanks,
and other large blow molded containers to ensure the structural integrity of these products whose performance is chal-

lenged by rigorous environments. The ability to achieve real-time, in-mold wall thickness measurement is desired by many manufacturers to allow for immediate adjustment of process parameters so that minimum wall thickness is ensured without compromising product integrity. Special ultrasonic sensors, conditioned for extreme heat and high pressure, are embedded within the walls of the mold. Measurement is taken of the sound pulse echo transmitted through the material. There are a some ultrasonic gauges currently available with this capability. However, the technique is still experimental and needs further development of fixturing and calibration compensation for temperature changes that result in velocity variations. Keep your eye out for major developments in this measurement area.

9.6 VISION SYSTEMS

Vision systems can be free standing or integrated on other on-line systems. They are comprised of a matrix that has vertical and horizontal cameras with on-line integrated SQC/SPC. Most systems require a lot of initial development work to ensure the correct settings for the specific application and to achieve desired tolerance, repeatability, and reliability. However, they are very reliable once up and running. Vision systems are available for a variety of uses including, but not limited to

> Dimensional gauging: top seal and surface ovality, cocked necks, T and E (see Fig. 9.1), outside diameter
> Label monitoring and location: voids, positioning
> Contamination and cosmetics: black spots, holes, bubbles, and other major deformities
> Shape recognition: pattern recognition, sorting
> Verification of completeness of subassembly: verification of tabs, welds, and clips
> Container sorting: identification and bar code reading

9.7   100% INSPECTION VERSUS RANDOM SAMPLING

Random defects can be effectively monitored measured using an on-line 100% inspection system, whereas key properties relating to process parameters are measured by random sampling. Random sampling identifies drift in process parameters so that corrective action can begin before quality drops below acceptable limits. Measurement data are available quickly and reliably allowing for continuous improvement and tight control over the manufacturing process. These initiatives improve quality and workmanship, including enhanced product strength, performance, and life span, which result in cost savings through reduced labor, lower material waste, and fewer customer claims, and rejections.

9.8   AUTOMATED VERSUS NONAUTOMATED
MEASUREMENT

Errors induced by operator variation, transposition, errors, and labor intensity of the measurement are inherent in any manual measurement. Most manual tests are subject to errors in operator technique. Three operators making the same exact measurements and using the same instrument on the same part will get three different results. This effect can be reduced by good training programs, or if practical, a system can be employed whereby one operator is responsible for the all measurements with a particular instrument or gauge. Manual measurements often require manual recording of results, which results in transpositional errors during the initial documentation and data entry. In addition, valuable time is lost while these measurements are being taken and the data are recorded. Although automated measurement instruments are more expensive, they should receive careful consideration because of the significant advantage they provide. Automated measurements are closer to real-time data and provide an early warning and quicker response time to process instability.

They provide immediate access to SPC/SQC data, and reduced measurement time translates into either larger statistically meaningful samples or more comprehensive testing. More importantly, they allow the quality technicians to analyze the data and interpret the results rather than spend all of their time taking the actual measurements.

9.9 QUALITY INFORMATION SYSTEM

Continuous improvement and the quest toward zero defects can be attained only if the quality control instruments, measurement data, management, production technicians, and quality technicians work as an integrated system. The main objective is to produce data to support monitoring and decision making for production optimization. In recent years computerized production control systems have been widely used. A system consists of modules or instrument groups. Located in the quality control laboratory or on the production floor (see Fig. 9.7). Instrument groups can operate as independent systems or as a part of a plant network. The goal is to make measurements and product data easy to enter, using touch screens, bar code entry and labeled keypads that require minimal use of the operator's time.

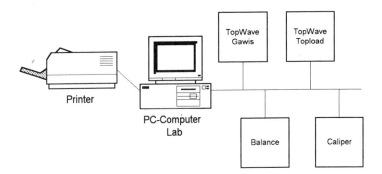

Figure 9.7 Instrument group.

A plant network consists of several instrument groups linked together in a standard local area network (see Fig. 9.8). This system allows, for instance, the laboratory instrument group to take individual measurements and generate QC charts, whereas the line instrument group interacts with QC charts and alarms for on-line and off-line measurements for this particular machine. The QC manager can review QC charts and oversee QC activity, while providing the Plant manger with a quality summary.

A wide area network (WAN) can be used to connect several plant networks with other networks allowing corporate headquarters access to data from various plants (see Fig. 9.9). This system is ideal for new product roll-outs where corporate quality and/or R&D can distribute specifications and measurement programs to ensure continuity through out the organization. Systems like these provide easy access to data, real-time data, automatic reporting, and homogeneous report formats.

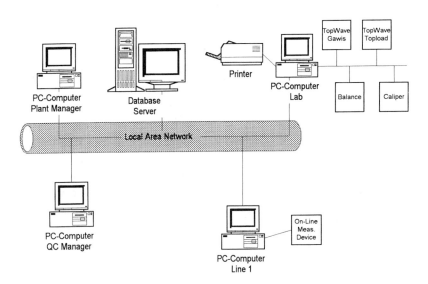

Figure 9.8   Plant network.

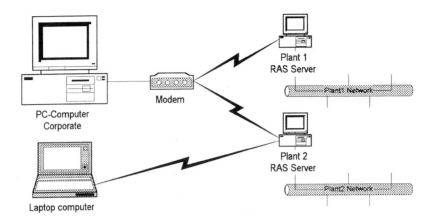

Figure 9.9   Wide area network.

9.10   CONCLUSION

It is impossible to cover in this chapter all of the details one should look at for testing extrusion blow molded products. Although sources are limited, numerous publications have been released that address areas of plastics testing. This chapter is only an overview of the many existing testing standards and procedures. Inquire of the many organizations, committees, and industry affiliates that commit them to establishing standards and test methods for a greater understanding of potential applications and continued growth of plastics (see Tables 9.4 and 9.5).

I can not stress enough the importance of a good testing plan implemented from part design, material selection, and incoming material inspection to process monitoring, quality control, and testing of the final product. It involves commitment from each segment of the organization and requires good communication among all parties, including the customer.

Table 9.4

National Standardization References

American National Standards Institute, Inc. (ANSI)	A nonprofit organization dedicated to the promotion of national standards, approval of standards at a national level, and avoidance of duplication of standards.
American Society for Testing and Materials (ASTM)	A society of technical, scientific, and educational members dedicated to promoting material engineering, standardization of specifications, and test methods (D-20, Technical committee on plastics).
Society of Plastics Industry (SPI)	A technical and trade society active in establishing test methods and specifications through its various divisions and membership committees.
Society of Plastic Engineers (SPE)	An international society, whose divisions represent each sector of the industry, dedicated to the progress of knowledge as it relates to the development, processing, and application of plastics.
Society of Automotive Engineers (SAE)	A society dedicated to promoting standards and engineering protocol connected with designing, constructing, and implementing self-propelled systems, their components, and related parts.
General Services Administration (GSA)	A government agency dedicated to establishing an efficient procurement system including development of federal specifications and standards covering mechanical, thermal, electrical, chemical, and physical tests. Standard #406.
National Bureau of Standards (NBS)	A cooperative program dedicated to promoting product standards for use in industry.

Table 9.5

ASTM Test Methods for Plastics

ASTM no.	Test method title
D-256	Impact Resistance
D-542	Index of Refraction
D-543	Resistance of Plastics to Chemical Reagents
D-568	Rate of Burning in Vertical Position
D-569	Flow Properties of Thermoplastics
D-570	Water Absorption by Plastics
D-618	Conditioning
D-621	Deformation of Plastics Under Load
D-635	Rate of Burning in Horizontal Position
D-638	Tensile Properties of Plastics
D-648	Deflection Under Flexural Load
D-673	Marring Resistance
D-695	Compressive Properties of Rigid Plastics
D-696	Coefficient of Linear Expansion
D-732	Shear Strength by Punch Tool
D-746	Brittleness Temperature by Impact
D-756	Weight and Shape Change Under Accelerated Conditions
D-785	Rockwell Hardness
D-789	Relative Viscosity, Melt Point, and Percent Polyamide
D-790	Flexural Strength and Modulus
D-792	Specific Gravity By Displacement Method
D-794	Permanent Effect of Heat on Plastics
D-952	Bond or Cohesive Strength
D-953	Bearing Strength of Plastics
D-1042	Dimensional Changes Under Accelerated Conditions
D-1043	Stiffness by Means of Torsion Test
D-1044	Resistance of Plastics to Abrasion
D-1180	Bursting Strength of Plastic Tubing
D-1203	Volatile Loss Using Activated Carbon
D-1204	Dimensional Changes at Elevated Temperatures
D-1238	Flow Rate by Extrusion Plastometer
D-1243	Dilute Solution Viscosity of PVC
D-1299	Shrinkage at Elevated Temperatures
D-1433	Rate of Burning at 45° Angle
D-1435	Outdoor Weathering of Plastics
D-1505	Density by Gradient Technique

Table 9.5 Continued

ASTM no.	Test method title
D-1525	Vicat Softening Temperature
D-1598	Time to Failure of Pipe under Pressure
D-1599	Short Term Hydraulic Failure/Plastic Pipe
D-1601	Dilute Solution Viscosity of Polyethylene
D-1603	Determination of Carbon Black in Olefins
D-1622	Apparent Density of Rigid Cellular Plastics
D-1623	Tensile or Adhesion of Cellular Plastics
D-1637	Tensile Heat Distortion Temperature
D-1652	Epoxy Content of Epoxy Resins
D-1693	Environmental Stress Cracking/PE
D-1708	Microtensile Properties of Plastics
D-1709	Impact by Free Falling Dart
D-1712	Resistance to Sulfide Staining
D-1790	Brittleness Temperature by Impact
D-1870	Elevated Temperature Aging in Tubular Oven
D-1893	Blocking
D-1894	Static and Kinetic Coefficient of Friction
D-1895	Density/Bulk Factor/Pourability
D-1925	Yellowness Index of Plastics
D-1938	Tear Propagation
D-1939	Residual Stress by Glacial Acetic Acid Immersion
D-2115	Oven Heat Stability of PVC Composites
D-2124	PVC Analysis by FTIR
D-2222	Methanol Extract of Vinyl Chloride Resins
D-2238	Absorbency of Polyethylene at 1378 CM-1
D-2240	Rubber Property Durometer Hardness
D-2288	Weight Loss of Plasticizer on Heating
D-2289	Tensile at High Speeds
D-2290	Tensile by Split Disk Method
D-2299	Determining Relative Stain Resistance
D-2383	Plasticizer Compatibility of PVC
D-2412	External Load by Parallel Plate Method
D-2445	Thermal Oxidative Stability
D-2463	Drop Impact of Blow Molded Containers
D-2471	Gel Time and Peak Exothermic Temperature Resins
D-2552	Environmental Stress Rupture, Constant Load
D-2561	Environmental Stress Crack, Containers
D-2562	Classifying Visual Defects in Thermosets

Table 9.5 Continued

ASTM no.	Test method title
D-2563	Classifying Visual Defects in Laminates
D-2578	Puncture Propagation Tear Resistance
D-2583	Barcol Hardness
D-2584	Ignition Loss of Cured Reinforced Resins
D-2587	Acetone Extraction and Ignition
D-2659	Column Crushing Thermoplastic Containers
D-2683	PE Socket Fitting Specifications
D-2684	Permeability of Containers to Reagents
D-2734	Void Content of Reinforced Plastics
D-2839	Use of a Melt Index Stand for Density
D-2841	Sampling of Hollow Mircospheres
D-2842	Water Absorption of Rigid Cellular Plastics
D-2856	Open Cell Content of Rigid Cellular plastics
D-2857	Diluted Solution Viscosity of Polymers
D-2863	Oxygen Index
D-2911	Dimensions and Tolerances of Plastic Bottles
D-2951	Resistance of PE to Stress Cracking
D-2990	Compressive Creep
D-2991	Stress Relaxation
D-3012	Thermal Oxidative Stability Biaxial Rotor
D-3014	Flame Ht., Burn Time, Wt. Loss in Vertical Position
D-3029	TUP Impact Resistance
D-3030	Volatile Matter of Vinyl Chloride Resins
D-3045	Heat Aging of Plastics Without Load
D-3124	Vinylidene Unsaturation Number by FTIR
D-3291	Compatibility of Plasticizer, Compression
D-3349	Absorption Coefficient, PE and Carbon Black
D-3417	Heat of Fusion by DSC
D-3418	Transition Temperature of Polymers by DTA
D-3536	Molecular Weight by GPC
D-3576	Cell Size of Rigid Cellular Plastics
D-3591	Logarithmic Viscosity of PVC
D-3593	Molecular Weight By GPC, Universal Calibration
D-3594	EA, EEA by FTIR
D-3713	Measuring Response to Small Flame Ignition
D-3748	Evaluating High Density Rigid Cellular Plastics
D-3810	Extinguishing Characteristics—Vertical Position
D-3835	Rheological Properties, Capillary Rheometry

Table 9.5 Continued

ASTM no.	Test method title
D-3846	In-Plane Shear of Reinforced Plastics
D-3895	Copper Induced Oxidative Induction by DSC
D-3914	In-Plane Shear of Rods
D-3916	Tensile Shear of Rods
D-3917	Dimensional Tolerance of Pultruded Shapes
D-3998	Pendulum Impact of Extrudates
D-4065	Dynamic Mechanical Properties of Plastics
D-4093	Photoelectric Birefringence
D-4100	Gravimetric Analysis of Smoke, Plastics
D-4218	Determination of Carbon Black Content
D-4272	Impact by Dart Drop
D-4274	Determination of Hydroxyl in Polyols
D-4321	Determination of Package Yield
D-4385	Classifying Visual Defects, Pultrudeds
D-4475	Horizontal Shear of Plastic Rods
D-4476	Flexural Properties, Pultruded Rods
D-4508	Chip Impact
D-4591	Determining Temperature Transitions by DSC
D-4603	Inherent Viscosity of PET
D-4660	TDI Isomer Content by Quant. FTIR
D-4662	Acid and Alkalinity Numbers of Polyols
D-4664	Freezing Point of TDI Mixtures
D-4665	Assay of Isocyanates
D-4666	Amine Equivalent of Isocyanates
D-4667	Acidity of TDI
D-4668	Quantification of Na and K in Polyols
D-4669	Specific Gravity of Polyols
D-4670	Determination of Suspended Matter, Polyols
D-4671	Determination of Unsaturation, Polyols
D-4672	Determination of Water in Polyols
D-4754	Extraction by FDA Migration Cell
D-4804	Flammability of Non-Rigid Solid Plastics
D-4812	Unnotched Izod Impact
D-4875	Quantification of PEO in Polyethers by NMR
D-4876	Acidity of Isocyanates
D-4877	Determination of Color, Isocyanates
D-4890	Gardner, Alpha Color of Polyols

ADDITIONAL READING

1. The Plastic Bottle Institute, Technical Bulletin PBI 21, The Society of the Plastics Industry, Washington, D.C., 1984.
2. The Plastic Bottle Institute, Technical Bulletin PBI 3 -1968, Rev. 2, The Society of the Plastics Industry, Washington, D.C., 1990.
3. The Plastic Bottle Institute, Technical Bulletin PBI 6 -1978, Rev. 2, The Society of the Plastics Industry, Washington, D.C., 1990.
4. The Plastic Bottle Institute, Technical Bulletin PBI 8 -1979, Rev. 1, The Society of the Plastics Industry, Washington, D.C., 1989.
5. D Rosato, D Rosato, Blow Molding Handbook, Hanser Oxford University Press, New York, 1988.
6. N Lee, Plastics Blow Molding Handbook, Van Nostrand Reinhold, New York, 1990.
7. A Pence, Smurfit Plastic Packaging, Wilmington, DE, interview-inspection and testing of drums.
8. R Enderby, Leak testing large blow molded parts, SPE Blow Molding RETEC, 1995.
9. ABC Group, Rexdale, Ontario, interviews-Automotive and industrial parts testing.
10. ABC Group, Rexdale, Ontario, Tech News 2(2), 1997.
11. H Mark, N Gaylord, N Bikales, Encyclopedia of Polymer Science and Technology, Wiley-Interscience, New York, 1970, Vol 7.
12. H Mark, N Gaylord, N Bikales, Encyclopedia of Polymer Science and Technology, Wiley-Interscience, New York, 1970, Vol. 8.
13. H Mark, N Gaylord, N Bikales, Encyclopedia of Polymer Science and Technology, Wiley-Interscience, New York, 1970, Vol. 9.
14. V Shah, Handbook of Plastic Testing Technology, pg. 202, Wiley-Interscience, New York, 1984.
15. NDT Applications No. 15, Wall Thickness Gaging in the Blow Molding Industry. Panametrics, Waltham, MA, Rev. 1996.
16. NDT Applications No. 21, In-Mold Thickness Measurement of Blow Molded Plastic Containers, Panametrics, Waltham, MA, 1991.
17. K Toivonen, TopWave Quality Control Network, TopWave International, Inc., Marietta, GA, 1994.

10

AUTOMATION, SIMPLICITY, AND PROFIT FOR BLOW MOLDING

Robert R. Jackson

Jackson Machinery, Inc.
Port Washington, Wisconsin

From initial approaches to standardization of parts to Henry Ford's first assembly line to today's flexible manufacturing cell, automation has been the vehicle to produce a quality product most economically. Each of these programs has had success along with some negatives, but automation programs continue to go forward each day. Blow molding, which is a relatively new manufacturing process, has been lagging in addressing automation. But now as the marketplace becomes more competitive, the need is more apparent and is a necessary step forward. We will try to address where and what the trends show.

Automation to reduce manufacturing costs is becoming a necessity. Eliminating labor content by outright reduction or by increasing operator output is not simple. Anyone who has tried automating a process without thoroughly investigating the process does not realize the value of machine operators and their ability to compensate for process variations and inconsistencies. It is an unspoken part of the job. The job may be boring and monotonous, but that operator is compensating for the problems that need to be addressed.

For automation to work requires repeatability of process and consistency of parts during the manufacture and the downstream utilization of the part. Consistency and repeatability usually require a better definition of product and upgrading of machinery and related process equipment and tools.

Blow molding also has the same potential and the same problems to remain competitive through the use of automation. The manufacturers of blow molded bottles have seen a more competitive environment for a number of years. The manufacturers of bottle type machinery using oriented discharge systems have had excellent automation available for fifteen years or more. Yet, many bottles are made on nonori-

Figure 10.1 SE-40 blow molding system with dual 6.8 kg (15 lb) accumulator heads.

ented discharge machinery. Blow molders using accumulator head type equipment are now seeing profit pressures and are addressing automation. That segment of the industry is in the early stages of this same automation revolution. However, molders are not buying automated accumulator head machines either.

This anomaly is the topic of our chapter. With the obvious aforementioned advantages of automation, why are we reticent to use it? The answer is that as an industry, we have not been forced by competition to become as creative as we could be. Who wants to spend an extra $100,000 to $500,000 on equipment when the older machines are still capable of providing marginal income? The reasons are quite complex.

Understanding of the cost of automation versus the cost of an employee is part of the key. A low skilled minimum wage employee's real cost is approximately $25,000 to $30,000 per employee per year. This takes into consideration the costs of a

Figure 10.2 Techne 1500 S, Bensenville, IL.

supervisor, workmen's compensation, insurance, and many other not readily apparent costs to the employer. The cost of operators could be generalized as double their rate of pay. In any case, this amounts to $75,000 to $90,000 per employee per year on a three-shift, five day per week basis. The machine that has two operators could have, in most cases, had none or one-half an operator doing packing. In this case automation could generate $100,000 to $150,000 a year of potential savings, plus any added savings from reduced scrap and greater throughput. However, this saving is not thought of as savings, but as out-of-pocket costs. The shortsightedness of this approach is one of the reasons that the blow molding industry is not progressing as rapidly as it could.

Another problem is that the setup of automatic machinery is thought to take more time than the manual machines and a higher level of skilled personnel. This objection is not really true. Using proper training and the right personnel allows automated machinery to show its real payback, even

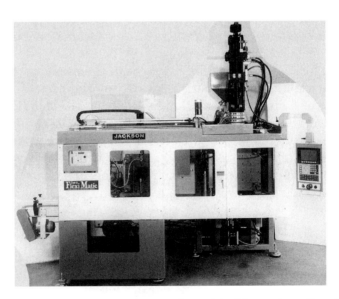

Figure 10.3 Fleximatic Type A, Jackson Machinery, Inc., Port Washington, WI.

with many mold changes. The next objection is the capital to buy the modern equipment. If the new machine costs $500,000 and the old would bring $75,000, the difference of $425,000 could be fully returned in two to three years.

Why will the blow molders not automate? More importantly, what will drive them to automation? First, the competition in blow molding is increasing. Secondly, the number of people to do this type of work is becoming increasingly small and finally, quality issues are becoming paramount.

A less expensive band-aid approach to automating bottle blow molding equipment is adding of takeout systems to older machines. For example, a takeout system that pulls the bottle into a deflashing station and then presents it for packing with oriented discharge can be tailored to most older machines. Other forms of automation, such as leak detecting, surface treating, spin off trimming and labeling, can all be done in-line once the container is oriented on a conveyor belt. There does not even seem to be a rush to do this. The fact that many

Figure 10.4 Fleximatic Type B, 5-lb head, Jackson Machinery, Inc., Port Washington, WI.

bottles do not need deflashing, just tail detabbing, offers some insight. In-mold labeling, on the other hand, cannot generally be added to older equipment but must be purchased as an integral part of the new machinery.

The very real savings that full automation with new equipment can give is still being missed even by large corporations. As molders approaches full automation, they and their customers' costs become minimized and the industry will operate at its most efficient levels. The folks who do this will stay in business, and the rest will slowly fade from the competitive business of making bottles.

The accumulator head machinery industry is not so fortunate in that the equipment that blows the part is fairly good, albeit a little inconsistent, but the in-machine trimming and oriented discharge are lacking. Automatic part removal is the simplest and singularly most cost effective improvement to manual blow molding machines. The reason for this is quite simply that the cycle time and consistency of that cycle time is now set by the machine instead of the operator. The importance of this cannot be emphasized enough. When operators set the cycle, part consistency, part quality, and part profitability are all affected by the randomness of the operator. Automatic part removing devices can be purchased from a number of generic suppliers and they come preengineered to interface mechanically and electronically with the machine. They range in price from $20,000 to $30,000 for machines from five pounds to fifty pounds in capacity. The use of strippers on smaller machines serves the same purpose.

Bottom blow stands are another method commonly employed to provide a finished or semifinished hole or neck in a particular part. These devices can be quite complicated and have parison spreaders and water-cooled blow pins for completely in-mold finished necks.

Semiautomatic automation in accumulator head machinery has been achieved by adding these secondary devices in a manufacturing cell concept. The operator acts as the transferring device, transferring the part from the machine or automatic part remover to a trimmer, a leak detector, or other secondary testing device and packs the part. These simple approaches to automation are increasingly used. However, they still do not allow full automatic part production. They are the same sort of add-on band-aids that takeout systems are to bottle machines. They improve the cost factors a little but do not go far enough.

In-mold trimming of the part is a relatively new technology available from some mold makers and replaces some of the aforementioned secondary operations. The downside of this method is the increased costs of the individual mold and the maintenance associated with the cutting devices in the mold. In-mold guillotining of part extremities is another method of automating part production. This method, as mentioned earlier, increases both the cost and maintenance of the mold.

Stand-alone trimmers are part of the manufacturing cell concept and are usually automatic once the part itself is placed in the machine. Once again, the operator is used as a transfer device. The trimmer can be hydraulically operated or air operated and has its own built in safeties for the operator.

A Leak detector can be used in conjunction with a stand-alone trimmer or as a separate device, once again using the operator as the transferring medium.

Each of the items is a band-aid fix. Instead of starting from scratch and doing it right, we are just delaying the inevitable noncompetitive situation on which our industry seems to thrive.

When questioned as a group or individually, the managers and owners of these blow molding shops all say they would love to throw out the older nonautomatic machines, but in most cases, their part volumes do not justify this expenditure. Again, shortsighted opinions. The new machinery that will work is flexible enough to allow for several different parts to be run, and the changeover time is rapid enough for the economies to make sense.

To achieve this, we have to start from scratch with new equipment and new thinking about how blow molding should be done. The extremely high production items, such as automotive fuel tanks have provided us with an idea of how this works. If we need new machinery because the old machinery cannot provide the needed product (multilayer fuel tanks), then we will fully automate the new machinery. When this new automated machinery is in place, we then look at the other older, nonautomated equipment and decide that maybe automation is an answer we should apply to all of our equipment.

Jackson Machinery, Kautex, and Hesta are several of the few companies that have recently introduced fully automatic in-machine trimming accumulator head type equipment. These are integrated systems that rely on the repeatability of the blow machine to allow the trim system to work well. The older machinery is not really capable of delivering the same part to the downstream trimmers each time. If the trimmer is constantly fed different sized parts, then it will not last very long. Thus, the new equipment has integrated in-machine trimming and also the latest state-of-the-art hydraulics and electronics which allow it to produce the same part time after time. This repeatability and consistency is the real key to true automation. Without it we are faced with another band-aid approach.

Another area of automation that causes tremendous problems is older molds. Those that were used in the manual machines are not well maintained or well designed and so do not work well with the automated machinery. When planning an automation project, we should be able to lower cycle times by at least 10%. This is achieved by a combination of consistent cycle times and new well-watered molds. The subject of well-watered, well-designed molds is a whole book in itself. It suffices to say that new molds are probably necessary to get the full value from automated machinery.

Blow molding automation requires quality molds and equipment that is repeatable from blown parts to finishing operations. It also requires flexibility to handle a range of parts to make it an economic reality. Today's proportional hydraulics and sophisticated controls are fully capable of providing the consistency needed. Now to the problem—money! $100,000 to $200,000 extra cost per machine. This requires a number of examples to justify fully automatic accumulator head blow molding machinery versus standard type equipment.

EXAMPLE I

A heavy handle ware bottle (approximately one gallon) is being run manually with two operators trimming, reaming, and blowing out the chips. Another part time employee wraps and ships. Absolute flatness is required on the top of the bottle.

The machine that runs this job manually can be bought new for $160,000 US and all the accessory manual trimming equipment for another $25,000 US, for a total of $185,000 US.

The automatic machine's price is $275,000 including the built-in trimmer. The bottle is switched from a blown finish to run in an invert mode with a blow pin, thus, ensuring a perfectly flat neck finish. The end user was willing to give the producer a five year contract when assured of perfect quality and a slightly lower price. The job now runs with no operators and one-half of a shipper, affording a three-shift, two-operator saving of $150,000 per year. Now, the automatic machine at the higher price is the best buy. The result was better pricing for the customer and increased profitability for the molder.

EXAMPLE II

The client was molding wheels in strings of three, with two accumulator heads, two machines and using one and a half people for each machine or three total per two machines. Production needs were in the $5.5 million range. The two manual machines cost $280,000 each with the trimming equipment adding another $50,000. The quality issues were that the occasional wheel with the center still in place caused final product failure. Two fully automatic machines with two strings of four up cost $375,000 each and require no operators. The parts have perfect hole centers by the nature of the process. The total invested capital in the manual machines and trimming equipment is $660,000. The fully automatic machine's cost is $750,000, and the labor saving is $225,000. The first year saving is $135,000 and better product. The production needs were exceeded which allowed for product growth.

EXAMPLE III

The producer was using five single-head machines to produce flashlight lantern bodies. This required five people per shift. The producer was running two shifts for a total of ten people. The old equipment cost approximately $75,000 each plus trimmers at $25,000 for each station. The total cost of the older equipment and trimmers was $500,000. The labor cost was $250,000 per year. The new fully automatic machine cost $250,000 and now supplies the final assembly line directly with no labor costs. The molds were replaced with new ones which greatly helped the quality issues and cost an additional $40,000. The new high technology machinery along with the new molds resulted in faster cycle times and therefore needed only two cavities. The old equipment was sold and paid for the molds. The labor saving alone paid for the whole project in the first year. The continued $250,000 per year saving went directly to the molder's bottom line.

If all this automation works so well, why is everyone in this business not doing it? There is this little problem of the extra $100,000–$200,000 initial cost per machine and the limited flexibility of the equipment to do multiple configuration parts. This extra invested capital gets into the molder's mind and spoils the pie. However, as stated earlier, the competition in getting good people and price pressure are beginning to change this.

The manufacture of fully automatic accumulator machinery requires a number of changes in core design to allow for the consistency necessary to make it work. The key word is consistency. To drop the same parison each time requires the highest form of machinery design. The Jackson machinery FlexiMatic™ blow molder is an example of the type of machinery that provides this consistency. It does it by incrementally small changes in each part of the machine, not by major changes in molding technology. For example, the heat exchanger is now controlled by a PID temperature zone. The viscosity of the oil determines the speed at which various cylinders operate, and even with proportional hydraulics, variation will occur if the oil temperature is not held constant. Feedback loops that proportionally control clamp speed, screw speed, push out, and 100 point parison programming, are all part of the overall requirements necessary to produce consistent parisons which produce consistent parts. When each part is not the same, the internal trimming system cannot survive. The sophistication of the control further helps track these variables for improved statistical process control. The Barber-Colman Maco 6500 is one of the more popular controls used in the blow molding industry. However, each machinery builder applies its own design to the control software. How this software is applied can make the control user-friendly or a nightmare. These changes also allow for the intricate machine adjustment criteria needed to enhance the consistency and quality of the parts produced. Further, the design of the system screens that interface with the set-up people and the operators can make a real difference in machinery performance. The control can also be networked to allow off-site troubleshooting and internal on-site information flow.

The profitability gains through automation can be tremendous. The machinery systems to accomplish these gains are available to all molders who are willing to look to the future. The move to full automation requires a very open-minded view of total profit and cost analysis. The underlying engine to these profitabilities will be found in the consistency of new age system components and designs. The cost is well worth the benefit. Don't be blind sided by your competition. Henry Ford is still right. Automation improves productivity and with full automation, we get even more cost reductions and quality improvements. Tomorrow's machinery is available now.

11

Troubleshooting Guide for Common Blow Molding Problems

Samuel L. Belcher

Sabel Plastechs, Inc.
Moscow, Ohio

11.1 PROBLEM SOLVING

Problem solving begins by ensuring that the resin used has the correct melt index and resin number and is free of impurities, metals, paper, and film. How many times have you had a blow molding line ready to begin production, only to find the wrong resin in at the machine, the wrong blend, or wrong color? It also begins with checking to be sure you have the correct packing material, labels, color number, and correct let down for the proper color match.

It is interesting to note that most major blow molders feed their machines from silos. Incoming resin is checked for melt index. However, the quality personnel are so busy or

Photo courtesy of Automa, Dublin, OH.

so staffed that the hopper car or hopper truck is already in the silo and mixed with resin already in the silo before they ever have time to run the melt index and report their findings to the purchasing or manufacturing department. Another problem exists when the inspection yields data showing that the melt index is not within the plant's specification, but it has already been unloaded, or the production department is told to use the resin. All of this requires that the production personnel have to change their machine settings to produce proper parisons for blow molding. With the movement to ISO 9000 and GMP (Good Manufacturing Practice), the use of out of specification resin should definitely be avoided.

Once the resin has been selected and approved, there are two other resin areas that should be checked and approved, the color, and the regrind.

Usually the color chip supplied by the color supplier is produced from virgin resin, and the color additive plus the color chip does not have the same thickness as the blow molded product. The color match should take into account the products thick and thin sections, it should be produced with the amount of regrind that will be used in actual production, and the carrier used for carrying the color additive should be virgin resin that has the same melt index and density. All too often the carrier is a higher melt flow material and possibly a low density polyethylene even though the part being produced is made of HDPE. You should know the exact specifications of your color additives. Because of the high cost of colorants, there is a large amount of mony to be saved by working with the colorant supplier. It is wise to have shear viscosity curves for the resin to be used in the product, for the resin with the amount of regrind being used, of the colorant itself, and finally, curves for the total blend of the resin, regrind, and colorant as it will run in production. All of the information should be sampled at the temperatures at which the part will be produced. This information should be transmitted to the production department so they know how the resin will flow and sag, the amount of die swell expected, the melt strength, and the surface appearance.

Part design should have been checked before ever trying to produce the product. Yet today many hours are spent redesigning the product at the machine.

Consistency in production allows for less labor, less downtime, less scrap, and it begins with the part design that allows consistent product based on consistent raw material supplied to the feed throat of the extruder.

Thus problem solving continues because the auxiliaries should all be checked for efficiency, accuracy, and full operation before starting production. The dryer, the grinder, the chiller, the blender, the thermolator, the reamer, deflasher, leak detector, labeler, box maker, and compressor (dry air) should be checked daily for each production line. How many times have you had a problem with production and you have to shut down because one of these items is not functioning?

Problem solving is continuous when in production. Even the simplest maintenance lapses affect cycle times and efficiency. Ensuring that everything on the machine and auxiliaries works and is properly maintained saves cycle time and provides production efficiencies. In many small and large production blow molding plants, the use of their off-fall or regrind can be the difference between profitability or loss.

Always check heater bands, thermocouples, limit switches, gauges, timers, mold alignment, cut-off knives, and hot wires for operation.

Remember if the units were all running satisfactorily before any problem occurs, don't panic. **THINK** before making any adjustments. Systematically make only one adjustment at a time and allow a minimum of twenty minutes for the adjustment to be effective.

11.2 COMMON TROUBLESHOOTING

11.2.1 Parison Problems

Problem	Possible cause	Suggested remedies
1) Doughnut formation	Cold mandrel	Allow mandrel to heat
	Mandrel extended too far	Adjust mandrel
2) Rough parison	Melt fracture	Increase melt temperature
		Adjust extrusion or parison drop rate
3) Parison sticking	Material or die too hot	Reduce appropriate temperature

4) Curtaining	High stock temperature	Reduce stock temperature
	Die misalignment	Realign die
5) Parison curl	Die alignment	Align die
	Loose mandrel	Tighten mandrel
6) Inconsistent parison lengths	Surging	Check extruder feed

11.2.2 Part-Related Problems

Problem	Possible cause	Suggested remedies
1) Poor welds and pinch-offs	Pinch area too hot	Decrease parison temperature
2) Thin parting lines	Insert misalignment	Increase pinch blade and align pinch-off insert
	Mold closing too fast	Decrease closing speed
	Poor alignment and/or mold venting	Align and vent properly
	Parison formation incorrect	Correct and ovalize bushing
3) Nonuniform wall thickness vertically, thinning from bottom to top. Circumferential nonuniform wall around part	Material too hot	Reduce material wall temperature
	Mandrel alignment	Align mandrel
	Die shaping incorrect	Reshape die
4) Poor surface	Melt fracture	Raise melt temperature
Rough surface		Decrease and/or increase parison drop rate
Streaks (die lines)	Burned polymer	Purge or clean die
	Contamination	Purge or clean die; also check feed of resins

Orange peel	Mold condensation	Increase mold temperature and/or cycle speed
Bubbles	Entrapped moisture and/or air	Check feed; possibly dry material
		Check for excessive fines
		Check feed rate; eliminate surge
		Check screw if continual problem
Contamination	Foreign particles	Check feed and material transfer section
		Check for degraded material in the head
Gloss	Parison too cold	Increase melt temperature
		Increase mold temperature and rate
		Increase blow pressure
5) Dimensional part weight too high or too low	Incorrect die opening	Adjust opening
	Parison programmer	Check parison program
Excessive flash	Excessive material	Increase melt temperature
	Swell	Reduce parison drop rate
Warpage	Poor cooling	Check chiller or thermolator; increase cooling; check mold for hot or cold spots; check for blocked water hoses or leaks
	Nonuniform material thickness	Adjust thickness

Excessive shrinkage	Part too hot	Reduce mold temperature; reduce material temperature increase cycle
6) Quality defects		
Missed handles	Misalignment	Align parison and die correctly
	Low swell	Increase parison drop rate; continually reduce temperature within operating range of extruder
Folds in the handle	Material too hot	Reduce melt temperature
	Curtaining	Reduce melt temperature
	Blow rate	Adjust preblow
Rings in the neck	Curtaining	Reduce melt temperature
	Blow rate	Decrease blow rate
Incomplete neck	Melt temperature too high	Reduce melt temperature
7) Excessive cycle	Melt temperature too high	Reduce melt temperature
	Mold temperature too high	Reduce mold temperature
	Wall thickness	Check die gap
	Improper mold design	Check water design
	Low blow pressure	Check air valves, air line, and compressor.
	Mold opening and closing	Check hydraulics and timers.

Any one problem or difficulty may have several causes so *think through the problem before making any changes.* Occasionally, changes in equipment design or end use requirements are necessary. A change in product design may be required to provide greater versatility and control.

Appendix A

MARLEX/PHILLIPS
POLYETHYLENE TECHNICAL INFORMATION
BULLETIN 31

Blow Molding:
Markets, Processes and Equipment;
Controlling Wall Distribution by
Programming and Die Shaping; Material
Requirements for Part Performance

INTRODUCTION

Blow Molding is the forming of a hollow object by inflating or blowing a thermoplastic molten tube called a "parison" in the shape of a mold cavity. Extrusion blow molding is the most widely used of several blow molding methods. The process consists of extruding or "dropping" a parison on which mold halves are closed. The ends of the parison are pinched shut by the closing mold, a pressured gas is introduced into the parison, blowing it out against the cavity walls (Figure 1).

FIGURE 1
Extrusion Blow Molding Process

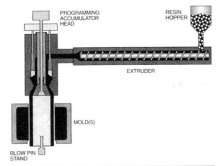

FIGURE 2
Blow Molding Cycle

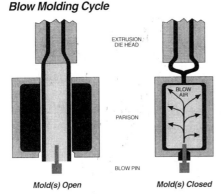

Mold(s) Open   Mold(s) Closed

The formed part cools (Figure 2), crystallizes and is further cooled until it can be removed from the mold without warping.

The second major blow molding method is injection blow, in which the parison(s) is injection molded onto one or multiple core rods and transferred to a blow mold(s) for part forming. A third method is oriented stretch-blow, a two step process in which the parisons are formed by either extrusion or injection molding. The cold parisons are reheated in a second operation, stretched (oriented) longitudinally and blow molded. There are more differences in blow molding equipment than for any other fabrication technique. A blow molding machine may be the size of an office desk, or may occupy a large room, making hollow objects as small as a pencil to 1300 gallons, or greater. There are also a large number of operating variables in a blow molding process which make it one of the more complex processes.

BLOW MOLDING MARKETS

SPI statistics show that 10.3 billion pounds of HDPE was consumed domestically in 1995 for all fabrication processes. Of this total, about 34% (3.5 billion) went into blow molding, making it the workhorse process of the HDPE industry. By comparison, 1987 HDPE blow molding usage was 2.7 billion pounds, and 1991 3.0 billion pounds.

The blow molding usage of 3.5 billion in 1995 was about 57% of the total 6.1 billion pounds of all materials that were blow molded. PET usage has increased rapidly in recent years and is now about 33% of the blow molding market. Polypropylene, PVC and LDPE make up the other 10% of the market. Blow molding consumption of materials such as K-Resin, engineering materials, elastomeric compounds, etc., were not reported by SPI.

Liquid food bottles consumed nearly 34% (1.18 billion lbs.) of the 3.5 billion total HDPE blow molding usage in 1995. Milk bottles were about 55% (0.65 billion) of liquid foods. Household chemical bottles (bleach, detergent bottles, etc.) were about 27% (0.92 billion) of the total poundage. About 73% of the 1995 blow molded HDPE resin went into bottles of all kinds, and the other 27% (0.94 billion) went into industrial items – gas tanks, drums, seats, ice chests, furniture, utility sheds, toys and other large items. The rate of growth has been faster in industrial parts than with bottles. In 1987, about 76% of the blow molded usage was in bottles, and 24% in industrial items.

PROCESSES AND EQUIPMENT

Blow molding can be classified in two major categories, extrusion and injection blow molding with many subdivisions. Continuous extrusion and intermittent extrusion are extrusion blow molding processes. Intermittent extrusion is further subdivided into reciprocating screw, ram, and accumulator parison extrusion. The second major category of injection blow molding "injects" the molten parison onto a metal core rod. Oriented bottle blow molding a combination of extrusion and/or injection processes. Molding equipment for these processes differ greatly. A brief description of each follows:

CONTINUOUS EXTRUSION BLOW MOLDING

A stationary extruder plasticizes and pushes molten polymer through the head to form a continuous parison (Figure 3).

FIGURE 3

Because the parison does not stop moving, it is necessary to "transfer" the parison from the die head to the mold(s) by means of arms, or by moving molds to the parison(s).

An example is a B-2 Kautex (Figure 4) once used in the Phillips 66 Plastics Technical Center (PTC). This machine transfers the parison away from the die head to a mold station located below.

FIGURE 4

The more common procedure with bottle molding is to move the molds to the parison. Bekum, (Figure 5) Kautex, Fisher, Hayssen, Johnson Controls and others build commercial continuous extrusion equipment which "shuttle" molds to the parison(s), at which time the press closes and captures the parison(s). Some manufacturers of large equipment are again using parison transfer in large part blow molding.

FIGURE 5

In Figure 5, the molds move diagonally up to the twin parisons being extruded. When the parisons are at the proper length, the mold halves close on the parisons. A hot or cold knife then cuts the parison between the top of the mold and die head. The press and molds move back down to the molding station (shown with the alternate press in molding position). Two calibrating blow pins move down inside the top of the parisons, compression molding neck seals. Air is introduced inside of the parisons, blowing the parisons out against the cavity walls, forming and cooling the part. The two presses with double cavity molds alternately move back and forth giving four-mold (4-up) production. In other machines, the press motion may be horizontal, an arc, or a straight downward movement to the parison(s).

Some of the biggest blow molders in this country use continuous extrusion, including Continental Can (also called Con Can and Continental Plastic Industries). They use large extruders and "wheels" which may hold up to 25 molds (Figure 6).

FIGURE 6
Continuous Tube Process – II

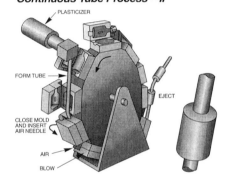

Some of the newer machines use side by side molds with twin parison extrusion. As the wheel rotates, the molds index and close on the parison. The machines use a needle "puncture" to introduce air inside of the parison. Rotary equipment is expensive. This is very high production rate equipment and requires millions of parts to be economical. Innopak (formerly Monsanto & "PLAX") also has this type of equipment. Graham Engineering also has a rotary machine – a vertical wheel, shown in Figure 7.

It extrudes the parison up, the parison is captured by the closing mold rotating up and away from the die head. Each mold on this machine also uses a needle blow.

FIGURE 7

INTERMITTENT EXTRUSION – RECIPROCATING SCREW MACHINES

Intermittent extrusion is often called "shot" extrusion. The screw rotates and retracts, plasticizing or melting the resin as it moves back, thus "charging" the shot in front of the screw. The screw is rammed forward by hydraulic means, pushing the plastic through the head and out the die head tooling as a parison (Figure 8).

FIGURE 8
Intermittent Parison Extrusion Blow Molding Using Reciprocating Screw

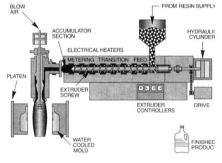

At this point, the mold(s) close. Parison extrusion is rapid, compared to a continuous extrusion. For a typical blow molding machine producing milk bottles, the parison extrusion time will be $1^1/_2$ seconds. A two head Uniloy dairy machine used by Phillips for studies, customer problems and operator training is shown in Figure 9.

It has the same controls as commercial Uniloy four- and six-head machines used to mold more than 90% of the milk bottles in the U.S.

Figure 10 shows the cooling conveyor table of a commercial dairy machine. In a normal milk bottle production line there are either four or six bottles on the cooling table, depending on the number of heads on the machine. The cooling table delivers the bottles with "flash" (flash is the part of the parison exterior to the cavity which is compression molded between mold halves) to the trimming conveyor.

These bottles are automatically indexed into the trimmer, cutting or knocking off the neck, handle, and bottom flash. The bottles are automatically leak tested in-line and conveyed to a silk screen printer where the appropriate dairy logo is printed on the milk bottle. From the decorating operation the bottles are conveyed via a combination conveyor/storage line prior to going to the milk filling equipment.

The conveyor system has a large amount of "surge" to keep the filling lines going in case the blow molding machine is shut down for short periods of time. In a normal milk bottle molding operation, the plastic bottle is not touched from the time it is blow molded until after it has been filled and capped.

The largest reciprocating screw blow molding machines sold commercially in the U.S. are the B-24, B-30 and B-50 IMPCO's with four, five and twelve pound parison shot extrusion capacities respectively. A B-24 is shown in Figure 11. These machines are used mostly for small and medium size industrial parts.

FIGURE 9

FIGURE 10

FIGURE 11

INTERMITTENT – RAM EXTRUSION, PISTON AND ACCUMULATOR HEAD

Figure 12 shows a schematic of an intermittent extrusion blow molding machine with a ram (piston) accumulator remote from the die head.

FIGURE 12
"Ram" Accumulator Blow Molding Machine

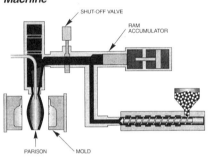

The ram pushes the accumulated material through the die head to form the parison as a "shot". Although this is a nearly obsolete process, this type of equipment is still operating in the U.S. and throughout the world. The extruder for this process is stationary and can continuously feed molten polymer into the accumulator. Or, the extruder can be programmed to stop when the die is filled to a pre-determined volume, and start again after the parison is extruded. The extruder speed is normally adjusted to fill the accumulator at the appropriate rate so that the parison is extruded almost immediately after the mold has opened and the part removed. An example of this kind of equipment is the old Producto shown in Figure 13.

Black Clawson, Hartig and others have a lot of this type of equipment still in operation; however, although the ram parison extrusion process generally has been replaced by accumulator heads. A pictorial of a Kautex last-in-first-out (LIFO) accumulator head is shown in Figure 14.

The LIFO head, in turn, has been replaced with first-in-first-out (FIFO) heads which have overlapping melt streams in the head to minimize weld (knit) lines and achieve stronger welds in the parison and part.

FIGURE 13

FIGURE 14
Kautex In-Line Head

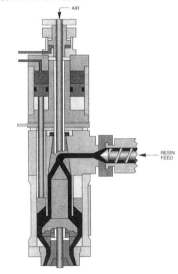

The accumulator head contains a complete parison shot in annular form. This results in a parison of much improved circumferential wall thickness uniformity, as well as a more uniform melt temperature. The parison is extruded by moving the annular ring downward, displacing the material through the die bushing and mandrel. The Hartig in the PTC has a 50 pound Kautex

FIFO accumulator head as shown in Figure 15. This machine is used to make fuel tanks, 30 gallon drums, large panels for utility sheds, garbage cans and other large parts.

FIGURE 15

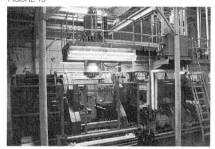

Hartig, Johnson Controls, Sterling Division of Davis Standard, Cincinnati Milacron and Techne-Graham are among the U.S. manufacturers of machines with accumulator heads which are similar in concept to the Kautex head, although all have significant differences. Bekum and Battenfeld are also major German manufacturers of large accumulator head machines. The largest blow molding machine in the world (as of 1977) was the Kautex shown in Figure 16.

Four extruders feeding an accumulator head with 300 pounds shot capacity is shown blow molding a 1320 gallon heating oil tank. The machine is located in Germany. Larger machines have been built since that time. Machines in this size range use high molecular weight resins so that the parison will have enough melt strength to "hang" on the die head during mold closure.

FIGURE 16

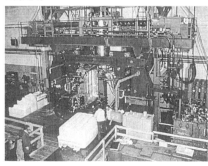

INJECTION BLOW MOLDING

An injection blow molding schematic is shown in Figure 17. The basic steps of injection blow molding comprise (1) injection molding melt around a core rod in an injection mold, (2) transferring the core rod and parison to a blow mold to form the part and (3) part removal. A PTC laboratory Jomar machine is shown in Figure 18.

FIGURE 17

Injection-Blow Process – I

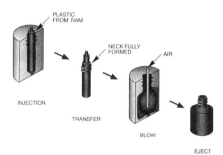

FIGURE 18

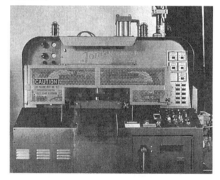

Normally the three stations of injection molding, blow molding, and part ejection are arranged 120 degrees apart in a vertical press with the core rods fixed on a horizontal indexing table for each station. The sequence is: injection mold the plastic on a set of core rods, the vertical press opens, the core rods rotate horizontally 120 degrees to the blow molds, the press closes, a new set of parisons are injection molded on the second set of core rods while the first set is being blow molded and the third station is ejecting the cooled parts off core rods. Figure 19 shows the bottom platen with core rods in the molding, blowing and ejection processes.

FIGURE 19

Advantages of injection blow molding are completely finished bottles with no deflashing step. This is an ideal process for pharmaceuticals, cosmetics, and toiletries. There are limitations to the process. Only bottles up to about one quart size are normally injection blow molded, and usually the parts are much smaller, with the sizes measured in ounces. To date, equipment has not been developed to make an item with a handle such as used with a one gallon milk bottle. Tooling is expensive and thus the process is best suited for high production volumes of small symmetrical bottles and vials. Rainville and Jomar are major manufacturers of commercial injection blow molding machines in the U.S. Wheaton, one of the largest HDPE users, also builds injection blow molding equipment, but only for their own captive molding operation.

ORIENTED PROCESS

There are several different concepts in oriented parison blow molding, one of which is the ORBET process developed by Phillips for polypropylene. By orienting a parison at a temperature slightly below the melting point of the polymer, significantly improved properties in terms of impact performance, stiffness modulus and clarity can be achieved. An ORBET orientation blow molding machine built by Beloit in the 1970s is shown in Figure 20. It is a two process operation.

FIGURE 20

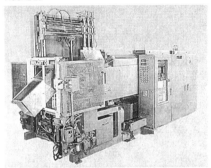

The parisons are made in an extrusion operation much like pipe. The cut-to-length open ended parisons are then reheated in an oven (Figure 21) to just below the melting point, transferred to the molding station (Figure 22) where they are stretched and oriented longitudinal by mechanical means (left view Figure 23). Axial orientation occurs when the parison is blow molded (right view Figure 23).

PET has replaced polypropylene as the preferred material in oriented-stretch blow molding because of both better barrier properties and clarity. An injection molded parison is used and the reheat characteristics of PET result in very high production efficiency, virtually obsoleting ORBET equipment.

FIGURE 21

FIGURE 22

FIGURE 23

CONTROLLING WALL DISTRIBUTION

Control of wall distribution is the heart of blow molding. Controlling wall distribution consists of choosing the correct parison size for blow molding a particular part, programming and die shaping.

The parison size (diameter) is very important – too small a parison results in a thick pinch-off which can cause warpage, and too high blow ratios to make the part without excessive thinning or "blowouts". Too large a parison can result in a thicker wall than wanted at the ends of the pinch-off, excessive "flashing", and trimming problems. The correct parison size is somewhere between these two extremes.

"Programming" is controlling the wall thickness of the parison and part, from top to bottom. The other primary technique of controlling wall thickness in a blow molding part is "Die Shaping" to thicken the parison and part in specific areas. Programming and Die Shaping are discussed in the following two sections.

PARISON PROGRAMMING, EXTRUSION BLOW

Figure 24 shows a programming die head. Programming thickens or thins the parison and part in the desired areas, as shown in Figure 25.

FIGURE 24

Moving Mandrel Programming – Electronic Control

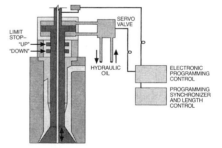

Programmed Parison Forming in a Mold Cavity

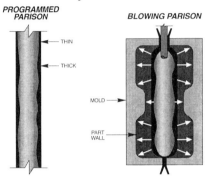

PROGRAMMED
PARISON   BLOWING PARISON

THIN

THICK

MOLD

PART
WALL

The center rod or "drawbar" is movable, by hydraulic means. The drawbar is normally attached to a "mandrel" that forms the inside of the parison.

A "bushing" forms the outside of the parison. The die head bushing and mandrel have either diverging (Figure 26) or converging lands.

The angular die lands result in die gap and resultant parison thickness changes with vertical movement of the mandrel. This is "parison programming".

Typical Bushing and Mandrel Configuration for Moving Die Programming

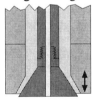

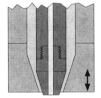

Diverging Land   *Converging Land*

Control of the programming is normally accomplished electronically, either with patch panels or micro-processor control. A patch panel is "programmed" with pins or slides for the amount of mandrel movement. Each point on the patch panel feeds a signal to a servo valve, which admits oil to a hydraulic cylinder at the top, resulting in drawbar and mandrel movement which is sensed by a linear variable differential transformer (LVDT). It feeds back a signal to the electronic programming patch panel to electrical null, at which point the servo valve shuts off oil, stopping movement. The programming proceeds to the next point on the patch panel. There may be 10, 20, 34 or more points on the programming panel, which means the same number of wall thickness changes can be made during a single parison extrusion. Most of the older electronic programmers had 34 points, or less. Figure 27 is a photograph of a Hunkar programmer on the large machine in the PTC. Two early manufacturers of patch panel programmers in this country are Hunkar and Moog.

Micro-processors have been used for programming in the past ten years or more. Micro-processors are more precise, have more programming points, and are more user friendly. The CRT screen can display the programming profile as values, bar graphs, or plots. Micro-processors do much more than program – they also monitor oil and water temperatures, control extruder/head temperatures, timers, etc. They are the brains of the blow molding machine. Barber-Coleman is a major supplier of micro-processors in the U.S., along with Hunkar, Moog, Siemens, Allen-Bradley and others. Some blow molding machinery manufacturers have their own micro-processor systems.

There are also mechanical means of programming by use of a rotary cam (Figure 28) to control or meter oil in and out of a programming cylinder.

FIGURE 28
Moving Die Bushing Programming

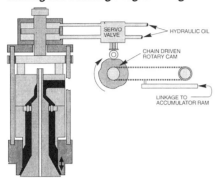

It is difficult to make rapid programming changes with a cam, and also difficult to synchronize the parison profile with part geometry. The cam system of controlling programming movement has been obsoleted by both patch panels and micro-processors.

In the illustration of Figure 28, the bushing is moved instead of the mandrel. The reason was to by-pass patents, which have now expired. The moving bushing head is much more massive than a moving mandrel head because drawbars and cylinders have to be located externally. Many moving die bushing heads are still in use, but all are controlled by either patch panels or micro-processors, rather than cams.

The Sterling Company (now a division of Davis-Standard) first introduced the micro-processor control for parison programming in their blow molding equipment line. They also controlled most of the machine sequencing with proportional hydraulics and micro-processors. All equipment manufacturers are now using micro-processor control systems for parison programming and sequence functions.

DIE SHAPING, EXTRUSION BLOW

The other primary technique often used to improve wall distribution on a blow molded part is die shaping. Die shaping, "profiling", or "ovalization" is machining the die gap larger in the appropriate sector to increase the parison thickness and part thickness where desired. Generally, the shaping cut is located in the exit land of the die. Shaping can also be placed above the land, but the cuts have to be deeper than if machined in the land. With many parts, shaping can also result in thinning areas which are too thick.

Irregularly shaped parts often require die shaping. Surprisingly, even round parts such as drums, and flat panel parts benefit from die shaping. Figure 29 shows the cross-section of a bushing with four shapes located in the appropriate position to increase corner thickness of a rectangular part.

Figure 30 shows a schematic of a shaped parison blowing into the corner of a box-like item.

FIGURE 29
Wall Distribution with Shaped Bushing

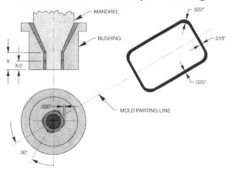

FIGURE 30
Shaped Parison Blowing into a Rectangular Mold Cavity

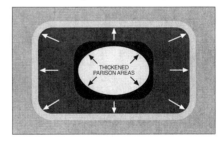

Choosing the areas to be shaped is generally done by marking the parison with a grease pencil as the parison is extruding. Computer calculated methods are being developed, but have not yet been refined for irregular shaped parts. By trial and error, the area or areas of the parison that require shaping are marked (Figure 31).

FIGURE 31

The corresponding positions are then marked on the die tooling. Each position is marked because flow patterns through the heads are not always symmetrical; therefore, calculated shaping positions would not necessarily correspond to the geometry of the part. A properly marked parison for a square shaped battery box is shown in Figure 32. The bottom of the die bushing with the four shapes for this part is shown in Figure 33.

FIGURE 32

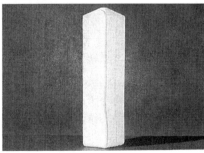

FIGURE 33

The depth, width and land length of the shaping cut in the die is based on experience. Thin wall gallon bottle tooling usually is shaped 0.002 – 0.003" deep. That's very little, and requires a precise set-up by the machinist (Figure 34).

On an intermediate thickness part, the shaping may be 0.060" to 0.080" deep and 0.150" to 0.180" for thick wall parts. Figure 35 is a photo of two intersecting shapes being cut in a 12" diameter bushing for a fuel tank.

FIGURE 34

The cuts are 0.080" and 0.120" deep, which is typical for this type of part. Depths of cut are generally 25% to 50% of the measured die gap, and 75% to 90% of the land length. It is a good practice to measure the minimum and maximum die gaps to help estimate the depth of the shaping cut. The width of the shaping cut varies greatly. Usually, the parison from wide shaping cuts blow more uniformly with less "thinning" at the edge of the shaping. Wide shaping cuts of nearly the tool diameter are often used with round parts. Shaping widths to place more material into corners are usually narrower, but still are often 50% to 70% of tool diameter.

FIGURE 35

Some round symmetrical parts such as a drum also require die shaping, because the areas of parison forming the top and bottom corners have to blow (stretch) the radius of the part 90° from the parting line. Correct shaping results in a smooth, thicker parison in the shaped area, which will cause few problems in blow molding (Figure 36).

FIGURE 36
Cross-Section of Shaped Parison

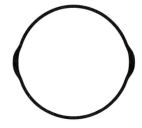

Shaping too deep will result in the parison extruding more rapidly and a longer parison in those areas, but the shaped parison areas are still a part of this slower moving, unshaped parison. The convoluted, unstable parison in Figure 37 is an example of too much shaping.

FIGURE 37

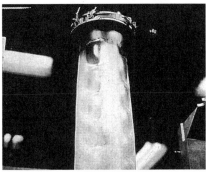

The solution is to machine out part of the shaping. It is easier to remove shaping than to shape deeper, as the machining set-up is concentric rather than eccentric.

Sometimes, a combination of programming and die shaping is desirable. This is the case in a round chair base where shaping is placed in a converging mandrel (Figure 38).

FIGURE 38
Programming Shaping into a Part

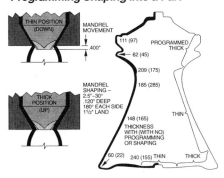

The die mandrel moves up to the maximum die gap and extrudes a thick, shaped parison into the 90 degree corners, where needed. At this position, the shaping is all within the die and maximum benefit from the shaping is achieved. As the die gap is reduced to thin the parison for the middle of the part, the shaped mandrel extends out of the bushing, reducing the effective shaped land length. This is effective with a cylindrical part because only a small amount of shaping is needed in the center of a round part. When the mold halves close, both ends of the parison are flattened, resulting in an elliptical or oval shaped parison. The areas near the end pinch-offs will blow to the wall first (since they are very close) and as a result be relatively thick. The corner areas 90° from the pinch-off are formed from parison that blows the radius of the part, requiring maximum programming and shaping.

MATERIAL REQUIREMENTS FOR GOOD PART PERFORMANCE

Material requirements for different part applications vary greatly. High stiffness, excellent stress cracking resistance (ESCR), extremely good impact properties and great processability are generally desirable in a blow molding resin. Resin development people constantly strive to develop one super resin having all these properties. Significant progress has been made in new catalyst technology resulting in better properties, as signified by the introduction of new resins from several suppliers. However, to date, there is no one resin that achieves all these goals. For this reason, resins are chosen for the need of a particular part—many blow molded applications don't need maximum performance of all four properties (high stiffness, impact, ESCR and processability).

Resin requirements fall into three general categories:

Short Life Applications (one year or less)

Short life applications generally use a resin that has moderate impact, and low stress cracking resistance. High stiffness is a must to minimize weight. Excellent processability is required to achieve high production rates. Typically, homopolymer resins that have a relatively high MI of about 0.6 to 0.8 are used to achieve high production rates. These nominal 0.7 MI resins will process at relatively low temperatures which reduces cooling time. These resins usually have a density in the 0.964 range for high stiffness. High

stiffness allows weights to be minimized and also increases production rates because of more bottles in a given output from the extruder, and less cooling time for the thinner wall. Milk bottles (Figure 39), water bottles and many other small parts fall into this grouping. Milk bottles alone consumed about 0.65 billion pounds of HDPE in 1995.

FIGURE 39

Moderate Life Applications (one to five years)

Moderate life applications use resins that have good impact and fair to excellent stress cracking resistance. To achieve up to five years life in an outdoor environment, pigmentation and addition of UV stabilizers in HDPE are usually necessary. Lowering the density (Flex Modulus) of moderate life resins increases environmental stress crack resistance (ESCR) of a resin made at the same molecular weight (MW) with the same catalyst system. These compromise resins, often called "general purpose resins", usually have MI's in the 0.1 to 0.4 range, with densities of 0.946 to 0.958. Even so, they are classified as high molecular weight resins (HMW). There is a wide range of melt index (MI), densities, and catalyst types used to make these resins, so significant differences can be found in resins used for moderate life usage. By far, the largest volume of blow molding resin usage falls in this category, with about 2.6 billion pounds per year blow molded into bottles and industrial parts in this grouping. Examples are the collage of bottles seen in any super market, as illustrated in Figure 40.

The one-gallon anti-freeze bottle in Figure 41 is another example of a part made with a general purpose resin.

FIGURE 40

FIGURE 41

Larger parts made from this type of resin include ice chests, water coolers, recreational items, assemblies of double wall panels into products such as rocking chairs, furniture, etc.

All of the following parts are made from Marlex® HHM 5502 or Marlex HHM 5202, or equivalents. These are 0.3 MI, 0.955 or 0.952 density resins, having moderate stiffness, acceptable ESCR for the applications, good impact and processability.

FIGURE 42
Ice chest on wheels

Rubbermaid

FIGURE 43
Work desk corner unit

(Corner King)

FIGURE 44
Five gallon rectangular water cooler

FIGURE 46
Five gallon round water cooler

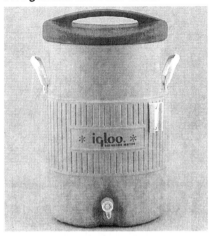

FIGURE 45
Steerable snow sled

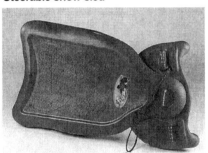

Rubbermaid SPD

FIGURE 47
File cabinet on wheels

Rubbermaid OPD

FIGURE 48
Rocking chair

The machines designed to run EHMW HDPE resins have screws and extruders designed for these materials, so melt temperature and outputs are optimized. Most parts using long life resins are large, but not all.

Examples of applications needing the superior properties of a EHMW resin include the parts shown in Figures 49 thru 58.

FIGURE 49
Stadium seats – Arrowhead, KC

Long Life Applications (greater than 5 years)

Long life applications use resins that have superior impact and enviromental stress crack resistance (ESCR) properties. Pigmentation and addition of UV stabilizers are also necessary in outdoor applications to achieve long life. The resins used in long life, high performance applications are "extra high molecular weight resins" (EHMW), usually with the melt flow measured as high load melt index (HLMI) – a flow too viscous to be measured by standard MI procedure. HLMI of these resins generally are between 2 and 15 – a big range. Densities of 0.946 to 0.958 are about the same as with the moderate life resins. The superior impact, toughness and impact properties are achieved because of the EHMW resin, combined with catalyst technology. The lower HLMI (high viscosity) resins can be difficult to process in conventional equipment, because shear heat can cause melt temperatures to override set points significantly. In equipment designed to handle EHMW resins, the high viscosity becomes an advantage with the parison having the high melt strength needed for long, heavy parisons.

FIGURE 50
Chemical containers

FIGURE 51
Jerry cans

Chilton Products

FIGURE 52
Chemical toilet, double wall, one-piece frame and door

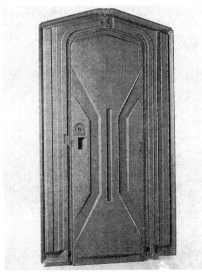

Satellite Industries

FIGURE 53
Pick-up truck tool box

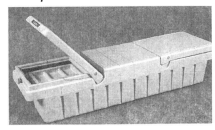

Delta, Inc.

FIGURE 54
Picnic table

Lifetime Products

FIGURE 55
Outboard fuel tank

Chilton Products

FIGURE 56
Escapement well for basement windows

R. M. Base

FIGURE 57
Fifty-five gallon drum

Resins used for stadium seats, chemical containers and Jerry cans are Marlex HHM 4903 and/or Marlex HXM 50100. Both are 0.949 density resins. HHM 4903 has an MI of 0.3 and HXM 50100 an HLMI of 10. Both materials have excellent ESCR. The tool box,

FIGURE 58
Automotive fuel tank

picnic table and the escapement well all use HXM 50100 for part toughness and excellent ESCR. The high melt strength of HXM 50100 is needed for the heavy, long parisons to make these parts. The five piece 66 inch wide escapement well has a total part weight of 131 pounds. This assembly forms the emergency escape stairs from a basement window. It must have excellent resistance to deformation due to earth loading, excellent toughness and ESCR.

Automotive fuel tanks are also made from Marlex HXM 50100 type of resins. An even tougher and higher ESCR resin, C579, is gaining acceptance as a superior fuel tank material. This type of material is a high EHMW resin with a HLMI of 5 and density of 0.944. The lower density greatly increases the toughness and stress cracking resistance.

"L-Ring" drums require great stiffness for compressive loads, superior impact at low temperatures, and excellent stress cracking resistance. The resins for these drums are Marlex C594 and TR570 types, with HLMI's in the 5 range, and densities in the 0.954 range.

These categories of resins vs. applications are not absolute – there are exceptions and crossovers of resins used in various applications for specific purposes, sometimes for the convenience of using only one resin in a plant. Pigmentation and UV stabilizers have to be used in HDPE for moderate and long life applications in outdoor environments. It is recommended that the resin supplier and processor work closely together in choosing a resin that best fits the needs, considering both part performance requirements and production efficiency.

Appendix B

MARLEX/PHILLIPS POLYETHYLENE TECHNICAL INFORMATION BULLETIN 32

Part 1. Blow Molding Highly Irregular Shaped Parts With Moving Mold Sections

Don L. Peters

Plastics Technical Center, Phillips Chemical Company, Bartlesville, Oklahoma

Part 2. Blow Molding Highly Irregular Shaped Parts With Multiple Parting Lines

Don L. Peters and J. R. Rathman

Plastics Technical Center, Phillips Chemical Company, Bartlesville, Oklahoma

Blow Molding Highly Irregular Shaped Parts With Moving Mold Sections

INTRODUCTION

Technology is presented disclosing methods of using moving mold sections to blow mold highly irregular part shapes in high density polyethylene. The significance of the technology is that it is the only known means to blow mold these parts in one piece. It is fairly common to move mold sections of a blow mold to release undercuts and permit part removal; however, the technique to affect or control wall thickness is used by relatively few molders. The ability to blow mold uniquely shaped parts has paid dividends by eliminating or reducing costly secondary operations. Three of the more important moving mold section techniques as developed at the Plastics Technical Center of Phillips Chemical Company are discussed.

INTEGRAL HANDLE LID

High density polyethylene water cooler lids with an integral handle have been blow molded commercially since 1965. Many millions of lids in three sizes have been produced using movable mold sections of a blow mold to form the handle. A lid is shown on a cooler in Figure 1.

Figure 1
Integral Handle Lid On Cooler.

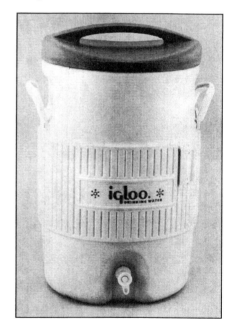

The mold consists of two basic mold halves - one half for the top of the lid and the other half for the bottom. The top half contains the movable "slides" to form the handles. The bottom half of the handle cavity is located in the movable slides and the top half in the stationary part of the mold. The horizontal slides are located under a face plate as shown schematically in the "open" position, Figure 2.

Figure 2
Integral Handle Mold Half Slides Open

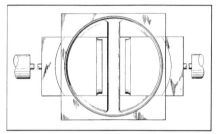

The molding sequence starts with a pre-pinched and pre-blown parison between the two open mold halves. The handle slides are open at this time. The mold halves are then closed on the parison. The closing action of the mold compresses and forces part of the parison into the handle cavity. A horizontal cut through the mold showing the parison "bubble" during mold closing is illustrated in Figure 3. A photograph of a production mold with the slides open is shown in Figure 4.

Figure 3
Integral Handle Mold Cross-Section Slides Open

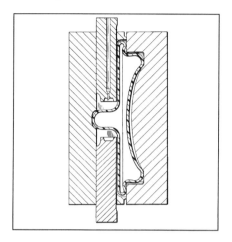

Figure 4
Commercial Mold – Handle Slides Open

3

The slides are then closed almost simultaneously with the mold closing. The blow air
is introduced through a needle into a flash
"pocket" under the handle. The blow air forms
the parison to the flash pocket shape, and
then flows through two open pinch blade
spots under the handle, resulting in the lid
body being blown through the handle passage
as illustrated in Figure 5. The handle slides
open with the mold opening to release the
undercut and allow the part to eject.

Processing efficiencies are excellent with precise control of the following variables: timing,
distance of slide opening, and speed of slide
movement. The amount of pre-blow air and
blow air timing also must be regulated and
synchronized accurately with respect to mold
closing.

The resultant lid has flash which must be
trimmed around the periphery and under the
handle; otherwise the handle is completely finished as the slides and stationary mold cavity
form the handle without flash. Good wall distribution is achieved as can be seen in the 3/4
section part photograph of a commercially
produced lid shown in Figure 6.

Figure 5
Integral Handle Mold Cross-Section – Slides Closed

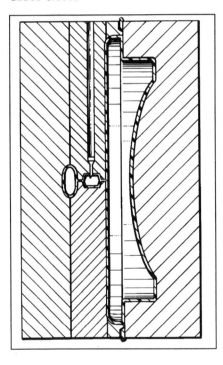

Figure 6
3/4 Section Of Lid

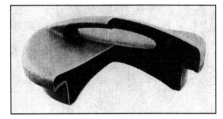

INTEGRAL HANDLE, DOUBLE WALL, INTERNALLY THREADED LID

Further development work resulted in a second high density polyethylene one-piece water cooler lid also having an integral handle plus an internally threaded double wall skirt. Hundreds of millions of these water cooler lids shown in Figure 7 have been produced commercially since 1968.

Figure 7
Integral Handle Double Wall Internally Threaded Lid On Cooler

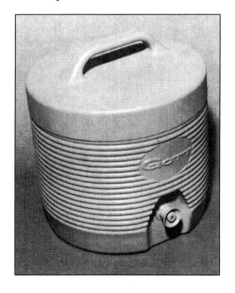

Again, there are two basic mold halves - one half to form the top and outside skirt and the other to form the internally threaded inside wall of the lid. The top half of this mold differs considerably from the first lid mold. This top mold half is split into two horizontally movable quarter mold sections to form the handle as shown in the upper part of the sectioned illustration, Figure 8. (This is in contrast to the first lid mold in which only the slide sections are movable.)

Figure 8
Open Quarter Mold Sections And Thread Forming Core

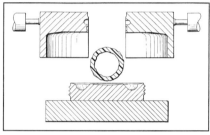

The bottom half is also quite different in that double wall blow molding technique is utilized to form the double wall skirt (lower illustration, Figure 8). A third distinction is the molding of the internal threads on the inside of the double wall. This necessitates an unscrewing core which releases the threaded lid off the core on mold opening.

A photo of a commercial mold with quarter mold sections open is shown in Figure 9.

Figure 9
Commercial Mold –
Split Quarter Mold Sections

The molding sequence requires that a pre-pinched and pre-blown parison be dropped between the mold halves. The quarter molds of the upper half are open at the time the parison is extruded. The press closes and almost simultaneously the quarter molds are closed by air or hydraulic cylinders to pinch the parison sector which forms the handle. The closing of the mold halves compresses and "balloons" the pre-pinched parison over the threaded core, "flashing" the cavity periphery.

Figure 10
Quarter Mold Sections Closed On
Thread Forming Core Half

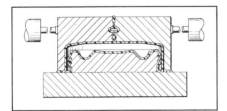

Blow air is introduced, forming the part as shown in the horizontally sectioned illustration, Figure 10.

At about the same time the press opens, the threaded core rotates to unscrew the part off the core. After a short delay, the quarter mold sections open. The delay on the quarter mold opening "holds" the lid so that the threaded core can unscrew rather than rotate the entire lid. Splitting the handle section of the mold into quarter sections results in the advantage of being able to mold a complex, irregularly shaped part with excellent wall distribution, as shown in the sectioned lid of Figure 11. The disadvantage is "flash" in two planes (Figure 12) which requires more trimming and regrinding.

Figure 11
3/4 Section Double Wall Lid

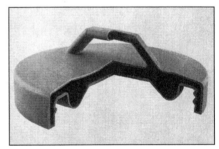

Figure 12
Lid With Flash In Two Planes

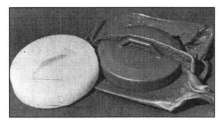

As the photo of the part shows in Figure 13, the lid can also be molded with internal threads in a compression molded solid wall. This type of lid has excellent thread strength, but it does not have the rigidity of a lid with a double wall skirt. It tends to warp more, and the outside appearance is not as good as the blow molded double wall part because of "sink" and "drag" marks.

Figure 13
3/4 Section Single Wall Lid

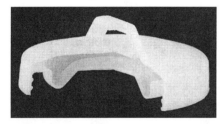

DRUM INTEGRAL HANDLING RING

A 30 gallon high density polyethylene drum with a handling ring integrally molded in the bottom chime area has been developed at the Plastics Technical Center in Bartlesville, OK. The handling ring allows metal drum lifting equipment to be used. A "double wall" relatively sharp radius chime was molded in the top (bung end) of the drum. The handling ring and sharp radius corner were located in their respective ends for mold building convenience and would be reversed in a production mold. A photo of the drum is shown in Figure 14.

Figure 14
30 Gallon Drum

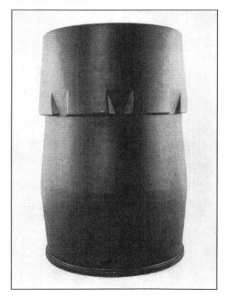

Although similar to the European "L Ring" drum in appearance and function, the Plastics Technical Center drum mold was designed and built independently. This drum mold has been used to develop molding techniques, to mold drums for testing (including handling), and for the evaluation of resins.

Each end of each mold half has a movable section, referred to as a "plug". The movable plug in the end of each mold half is semicircular, as shown in Figure 15. The open and closed positions of the plugs during molding are shown by the "dry cycle" photographs of one mold half (Figures 15 and 16, respectively). The mating plugs of two closed mold halves form the head and bottom of the drum mold. The plugs are moved by means of hydraulic cylinders on each end and both sides of each mold half (Figure 17). The 3-1/2" diameter hydraulic cylinders are securely mounted to the mold body and plug assemblies. Large cylinder rods (2-1/2" diameter) are used to assure smooth plug movement with no "cocking", caused by possible variations of hydraulic flow or pressure in the cylinders. Good wear plates are required between the plugs and mold body to allow plug movement under machine clamp and blow air pressures.

Figure 15
Plug Open Position

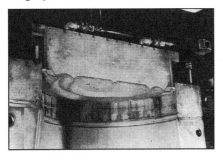

Figure 16
Plug Closed Position

Figure 17
30 Gallon Drum Mold

As is normal in large drum blow molding, the mold is placed in the press with the bung openings down. The molding sequence starts by extruding a parison between the mold halves (Figure 17). The parison is dropped over the two bung opening blow pins which serve as a parison spreader device. The blow pins are in the "together" position when the parison is extruded, and then move outward to a position corresponding to the bung openings in the mold halves. The mold halves are closed on the blow pins, either (1) compression molding bung threads on the outside diameter of the stub neck, or as in this case, (2) the blow pins hold injection molded high density polyethylene bung inserts which fusion weld to the parison forming the drum head.

During mold closing, the plugs are open (extended). With the plug ends open, low pressure blow air is introduced forming the parison to the cavity shape. A short time after the blow air is introduced, the plugs are closed (retracted) about 2.9". The extra length of parison is compression molded into the handling ring for the drum bottom. On the bung end, the extra parison length is moved into the drum top corners. The distance of plug travel is extremely important. Increasing the plug travel from 1.9" to 2.9" increased the wall thickness and improved wall thickness uniformity in both the handling ring and blown top corners. High pressure blow air is then admitted into the drum and the part cooled. The plugs retract on mold opening to release the undercuts created by the handling ring and top corner reverse radii. Separate timing was required for the plug movement on the bung end relative to the timing for the handling ring plug movement.

A combination of die shaping, parison programming rate and distance of plug travel results in a solid handling ring with uniform circumferential thickness as shown in the lower part of a drum cross-section, Figure 18. The same photo shows the thick corners achieved in the top corners of the drum.

Figure 18
Drum Cross-Section

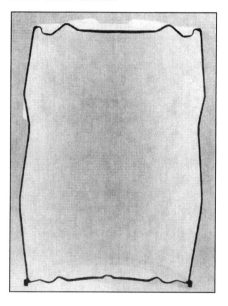

Reproducing wall distribution from part to part requires high precision timing of the blow air with respect to the plug movement. The handling ring can be made solid, semi-solid or hollow simply by changing the timing of the plug movement. Early plug movement results in hollow handling rings and later plug movement for solid rings (Figure 19). Although the plug timing is an easily adjustable process to control, it is also a process variable as normal operating "drifts" can cause the ring to vary in wall thickness from solid to semi-hollow, etc. The work at the Plastics Technical Center indicated that a solid ring can probably be molded consistently with good equipment, careful monitoring of the molding conditions and good quality control; however, molding a hollow ring with acceptable reproducibility is more difficult and possibly would require more precise controls and/or more consistent melt viscosity in resins than presently available.

Figure 19
Handling Rings Cross-Section

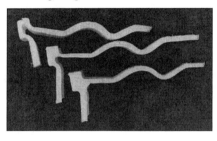

The same plug movement on the bung end of the drum resulted in good wall thickness in a sharp radius chime (no compression molding). This is significant in that drums with straight walls and sharper radius corners can be designed for good top load strength.

Drums molded with solid handling rings can be lifted with conventional drum lifting equipment; however, the hollow ring does not have enough wall thickness to withstand the frequently used "parrots beak".

High density polyethylene resins with about a 10 HLMI* have very good drop impact performance in conventional drums and are the standard material used in the United States. These same resins fail dismally in the solid handling ring drum. The poor impact results from a combination of three factors:

1) There usually is a slight "vee" notch between either the drum head and solid ring, or the drum side wall and the solid lid ring.

2) The wall thickness in the "vee" often is slightly less than the nominal wall thickness of the drum wall.

3) The slower cooling of the thick wall handling ring adjacent to the faster cooling walls probably results in molded-in stresses.

Poor drop impact properties of drums with solid rings molded with nominal 10 HLMI can be overcome by using higher molecular weight resins. HLMI versus drop impact of 30 gallon drums filled with water shows the effect of molecular weight:

HLMI	DROP IMPACT
10.0	2.5 ft.
2.2	7.2 ft.
1.9	9.3 ft.

At one time, high density polyethylene resins with a HLMI of about 2.0 were considered optimum for drums with solid handling rings. These resins were available only in fluff (powder) form. In recent years, 5 HLMI resins in pellet form have been found to perform very well in drum applications. Pelleted resins are generally preferred in the U.S.

* HLMI - High load melt index as measured by ASTM D1238 test method.

CONCLUSION

Many millions of water cooler lids produced in the past 29 years have proven the practicality and economics of using moving mold sections to blow mold irregularly shaped one piece parts otherwise not blow moldable. The technology should be applicable to other parts now made in two or more pieces, and to improve wall distribution in some blow molded parts produced conventionally.

Development work has shown that drums with an integral handling ring which can be handled with conventional lifting equipment can be blow molded in one piece using moving plugs. One of the most important keys to achieving desired wall thickness is the distance of plug travel. The molding process requires good equipment, careful monitoring of production, and high molecular weight high density polyethylene "fluff" resins with great inherent toughness. Plastics Technical Center work on blow molded drums with solid handling rings generally agrees with the known parameters used in molding the European "L Ring" drums. In the past ten years, 5 HLMI resins in pellet form have been found to process easier and have excellent performance. The 5 HLMI resins have replaced the 2 HLMI resins in most applications.

The ability to mold relatively sharp radii corners with good wall thickness using moving plug technique is significant. The resulting greater design freedom should improve part geometries and spawn new applications, particularly in the medium to large size parts. The higher precision of micro-processor controls on blow molding equipment should result in moving section blow molding becoming relatively routine.

References:

Patents assigned to Phillips Petroleum Company
U.S. Patent No. 3,342,916
U.S. Patent No. 3,424,829
U.S. Patent No. 3,438,538
U.S. Patent No. 3,585,681
U.S. Patent No. 3,792,143

BLOW MOLDING HIGHLY IRREGULAR SHAPED PARTSWITH MULTIPLE PARTING LINES

SUMMARY

Phillips Chemical Company has long been involved in the development of moving section molds to blow mold parts that otherwise could not be blow molded. Two basic moving sections molding methods (Patents 1, 2, 3, 4, & 5) have resulted in several hundred million water cooler lids being produced by two companies since conception. This work was originally reported at the 1982 ANTEC.

New multiple parting line technology has been developed which allows the blow molding of part shapes not possible with previous moving sections methodology. Articles with a reverse "fold" or inwardly protruding channel at the parting line of the mold can now be blow molded. Similarly, parts with narrow, double or single wall projections on either side of the parting line can be made with good wall distribution. The technology has been successfully demonstrated on a large mold at the Plastics Technical Center of Phillips Chemical Company. Disclosure of the work follows.

BACKGROUND

It was desired by a company to blow mold a one piece compartment separator which would replace an existing eighteen piece combination metal/ABS part. The flat, planar part required "legs" to protrude from each edge of both sides of one parting line side. The general configuration of the item is shown in Figure 1. Overall dimensions were about 29" x 23" x 2-1/2" thick. A mold was built to blow mold this shape using a conventional moving section to try to form the reverse fold. The moving member was attached to one mold half as shown in the cross-section, Figure 2. The molding sequence was to close the mold halves on a pre-blown parison with the moving

Figure 1

Desired Part Configuration

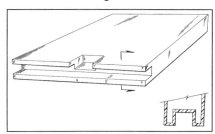

member (bar) in the retracted position. The positive pressure caused the parison to "flash" between one mold half and the movable member, resulting in good wall thickness in that leg. Blow air was introduced and the bar moved inwardly in an attempt to form the second leg. As the movable bar is fastened to this mold half, plastic must blow into the narrow width leg between the bar and mold face.

Figure 2

Mold With Movable Member Attached To Mold Half

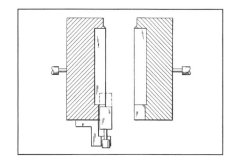

Figure 3
Ruptured Corners Encountered With Moving Member Attached To One Mold Half

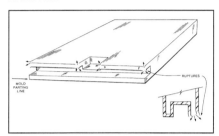

A complete part was not made by a custom molder in 1-1/2 years of trials. Blow-outs or ruptures always occurred in corners of the leg formed between the moving member and that mold half (Figure 3). Nominal wall thickness of 1/2 inch and part weights as high as 17 lbs. still resulted in ruptured corners, as shown in the photographed section, Figure 4. At this time the mold was evaluated at the Plastics Technical Center. Although much better parts were made at half the weight, ruptured or extremely thin corners still occurred.

Figure 4
Section Of Part Showing Ruptured Corner

THE NEW TECHNOLOGY

The new concept was simply to detach the moving member from its location on the one mold half and locate it in a stationary position midway between mold halves. This made the moving member a piece which was separate from the mold and it shall now be called the third member. The third member was mounted to a beam which was affixed to the press frame. This arrangement resulted in a second parting line, one on each side of the third member. A frontal view of the multiple parting line mold and separately mounted third member is shown in Figure 5. Pre-pinch plates are attached to the bottom of this mold.

Figure 5
Modified Mold With Third Member Mounted Between Mold Halves

Figure 6
Parison Extruded Adjacent To Third Member

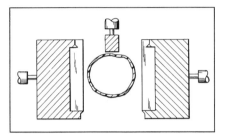

The molding sequence consists of extruding a parison between the mold halves and adjacent to the third member (Figure 6). The parison is then pre-pinched and preblown while the mold halves are open (Figure 7). This results in the parison starting to fold or wrap around both sides of the third member. Because there is open space between both sides of the third member and the mold halves, the flow of the pre-blown parison around the member is unimpeded. This is the principle design feature of this technology.

Figure 7
Pre-Pinched And Pre-Blown Parison Starts To Fold Around Third Member

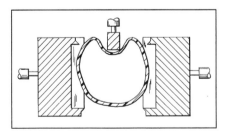

Figure 8
Pre-Pinched And Pre-Blown Parison Folding Around Third Member On Mold Closing

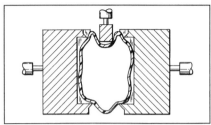

The closing action of the mold halves further compresses the trapped air inside the parison, forcing the molten plastic tube around both sides plus the top and bottom of the bar (Figure 8). The mold halves have pinch blades on the sides, top and bottom of the third member or mating mold faces. Adequate material is trapped around the third member to form legs having relatively uniform thickness (Figure 9). The wall thickness of the legs can be controlled to some extent by the amount of "wrap" or folding of the parison around the bar. A compartment separator made using the new technology is shown in Figure 10.

Figure 9
Mold Closes And Traps Pre-Blown Parison Between Third Member And Mold Halves

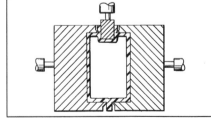

Figure 10
Complete Separator Made After Mold Modification

The parts made using this technology had a weight of 5.0 pounds and a nominal wall thickness of 0.100 inches. A series of parts were made under nominal production conditions. Good wall thickness distribution resulting from use of this methodology is shown in the photograph taken of a double leg cross-section, Figure 11.

Figure 11
Cross-Sections Of Separator Showing Wall Distribution In Double Legs

Reciprocating the third member is an option which would be desirable in deep undercut parts. This was evaluated and found unnecessary for this particular part.

CONCLUSIONS

Use of multiple parting lines in a mold closing on a stationary or reciprocating third member has resulted in the production of a part not possible by any other blow molding means. Figure 10 shows a photograph of a compartment separator made by this technology. Blow molding this part was proven as a viable technique in extensive runs at the Phillips Plastics Technical Center.

Figure 12
Using Multiple Parting Lines To Achieve Good Wall Distribution

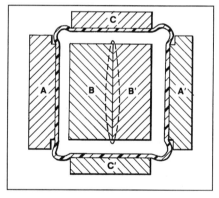

Extension of this development will allow the blow molding of box-like parts with sharp corners. An example is shown in Figure 12 where all mold sides and ends are split, resulting in a parting line at each edge. The parison is pre-blown until all parting lines are flashed,

Figure 13
Example Of Box Shape Moldable By Use Of Multiple Parting Line (Before Trimming)

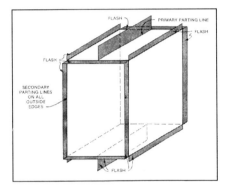

Figure 14
Example Of Part With Double And Single Wall Edges That Can Be Made With Multiple Parting Lines

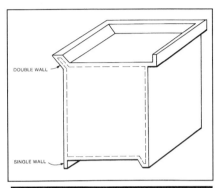

the mold members close, pinch welding at the edges and the part blow molded (Figure 13). The part must be trimmed at each pinch weld edge.

The same technology would allow the molding of double or single wall ears, legs or other projections at parting lines (Figure 14).

Parts made by rotational molding or two piece assemblies can now be considered for blow molding using this technology.

References

(1) US Patent No. 3,342,916
(2) US Patent No. 3,424,829
(3) US Patent No. 3,438,538
(4) US Patent No. 3,585,681
(5) US Patent No. 3,792,143
(6) Blow Molding Highly Irregular Shaped Parts With Moving Mold Sections - D. L. Peters

Glossary

Plastics Terms

The *Glossary of Plastics Terms* had its beginning over a decade ago. Originally, it was simply an unpublished group of terms used by Phillips engineers in their work on Marlex® plastics. The Glossary was actually published for the first time in 1959, as a convenience to Phillips customers and other friends interested in plastics. To make it more useful, the first edition included a number of terms which had appeared in MODERN PLASTICS ENCYCLOPEDIA, through the courtesy of its publishers. This practice has been continued in subsequent editions.

As the acceptance of the Glossary has grown, so has its size. The fourth edition includes a substantial additional number of terms from a variety of sources, including ASTM D883 – *Standard Definitions of Terms Relating to Plastics* . . . . making it even more interesting, useful and helpful than previous editions.

PHILLIPS CHEMICAL COMPANY
A SUBSIDIARY OF PHILLIPS PETROLEUM COMPANY

BARTLESVILLE, OKLAHOMA 74004

A

A-STAGE

An early stage in the reaction of a thermosetting resin in which the material is still soluble in certain liquids and fusible. See also *B-* and *C-Stage.*

ABLATIVE PLASTICS

This description applies to a material which absorbs heat (while part of it is being consumed by heat) through a decomposition process known as pyrolysis, which takes place in the near surface layer exposed to heat.

ABSOLUTE VISCOSITY

Of a fluid, the tangential force on unit area of either of two parallel planes at unit distance apart when the space between the planes is filled with the fluid in question and one of the planes moves with unit differential velocity in its own plane.

The C.G.S. unit for absolute (or dynamic) viscosity is the poise (dyne-sec./sq. cm.). The centipoise (0.01 poise) is often used.

ACCELERATOR

A substance that hastens a reaction, particularly one which speeds up the vulcanization of rubber. Also known as *Promoter.*

ACCUMULATOR

A term used mainly with reference to blow molding equipment which designates an. auxiliary ram extruder which is used to provide extremely fast parison delivery. The accumulator cylinder is filled with plasticated melt coming from the extruder between parison deliveries or "shots" and is stored or "accumulated" until the plunger is required to deliver the next parison.

ACETAL RESINS

The molecular structure of the polymer is that of a linear acetal, consisting of unbranched polyoxymethylene chains.

ACRYLIC ESTER

An ester of acrylic acid, or of a structural derivative of acrylic acid, e.g., methyl methacrylate.

ACRYLIC RESIN

A synthetic resin prepared from acrylic acid or from a derivative of acrylic acid.

ACRYLONITRILE

A monomer with the structure $(CH_2:CHCN)$. It is most useful in copolymers. Its copolymer with butadiene is nitrile rubber, and several copolymers with styrene exist that are tougher than polystyrene. It is also used as a synthetic fiber and as a chemical intermediate.

ACRYLONTRILE – BUTADIENE-STYRENE (abbreviated ABS)

Acyrlonitrile and styrene liquids and butadiene gas are polymerized together in a variety of ratios to produce the family of ABS resins.

ADDITION POLYMERIZATION

Polymerization in which monomers are linked together without the splitting off of water or other simple molecules.

ADIABATIC

An adjective used to describe a process or transformation in which no heat is added to or allowed to escape from the system under consideration. It is used, somewhat incorrectly, to describe a mode of extrusion in which no external heat is added to the extruder although heat may be removed by cooling to keep the output temperature of the melt passing through the extruder constant. The heat input in such a process is developed by the screw as its mechanical energy is converted to thermal energy.

ADHESION PROMOTER

A coating which is applied to the substrate before it is extrusion coated with the plastic and which improves the adhesion of the plastic to the substrate.

ADHESIVE

A substance which applied as an intermediate is capable of holding materials together by surface attachment.

ADSORPTION

The adhesion of the molecules of gases, dissolved substances, or liquids in more or less concentrated form to the surfaces of solids or liquids with which they are in contact.

AFFINITY

The attraction for another substance.

AGING

The change of a material with time under defined environmental conditions, leading to improvement or deterioration of properties.

AIR-ASSIST FORMING

A method of thermoforming q.v., in which air flow or air pressure is employed to partially preform the sheet immediately prior to the final pulldown onto the mold using vacuum.

AIR GAP

In extrusion coating, the distance from the die opening to the nip formed by the pressure roll and the chill roll.

AIR RING

A circular manifold used to distribute an even flow of the cooling medium, air, onto a hollow tubular form passing through the center of the ring. In blown tubing, the air cools the tubing uniformly to provide uniform film thickness.

AIR-SLIP FORMING

A variation of snap-back forming in which the male mold is enclosed in a box in such a way that when the mold moves forward toward the hot plastic, air is trapped between the mold and the plastic sheet. As the mold advances, the plastic is kept away from it by the air cushion formed as described above, until the full travel of the mold is reached, at which point a vacuum is applied, destroying the cushion and forming the part against the plug.

ALIPHATIC HYDROCARBONS

Saturated, hydrocarbons having an open chain structure. Familiar examples: gasoline and propane.

ALKYD RESIN

Polyester resins made with some fatty acid as a modifier. See *Polyester, Fatty Acid.*

ALKYL

A general term for monovalent aliphatic hydrocarbon radicals.

ALLOY

Composite material made up by blending polymers or copolymers with other polymers or elastomers under selected conditions, e.g., styrene-acrylonitrile copolymer resins blended with butadiene-acrylonitrile rubbers.

ALLYL RESIN

A synthetic resin formed by the polymerization of chemical compounds containing the group $CH_2{=}CH{-}CH_2{-}$. The principal commercial allyl resin is a casting material that yields allyl carbonate polymer.

ALPHA-CELLULOSE

A very pure cellulose prepared by special chemical treatment.

AMINO

Indicates the presence of an $-NH_2$ or $-NH$ group.

AMORPHOUS PHASE

Devoid of crystallinity — no definite order. At processing temperatures, the plastic is normally in the amorphous state.

ANGLE PRESS

A hydraulic molding press equipped with horizontal and vertical rams, and specially designed for the production of complex moldings containing deep undercuts.

ANILINE

$C_6H_5NH_2$. An important organic base made by reacting chlorobenzene with aqueous ammonia in the presence of a catalyst. It is used in the production of aniline formaldehyde resins, q.v., and in the manufacture of certain rubber accelerators and antioxidants.

ANILINE FORMALDEHYDE RESINS

Members of the aminoplastics family made by the condensation of formaldehyde and aniline in an acid solution. The resins are thermoplastic and are used to a limited extent in the production of molded and laminated insulating materials. Products made from these resins have high dielectric strength and good chemical resistance.

ANNEALING

A process of holding a material at a temperature near, but below, its melting point, the objective being to permit stress relaxation without distortion of shape. It is often used on molded articles to relieve stresses set up by flow into the mold.

ANTI-FRICTION COMPOUNDS

Materials specifically formulated to reduce or eliminate friction.

ANTIOXIDANT

Substance which prevents or slows down oxidation of material exposed to air.

ANTISTATIC AGENTS

Methods of minimizing static electricity in plastics materials. Such agents are of two basic types: (1) metallic devices which come into contact with the plastics and conduct the static to earth. Such devices give complete neutralization at the time, but because they do not modify the surface of the material it can become prone to further static during subsequent handling; (2) chemical additives which, mixed with the compound during processing, give a reasonable degree of protection to the finished products.

ARC RESISTANCE

Time required for a given electrical current to render the surface of a material conductive because of carbonization by the arc flame.

ARMOR

A solid or braided metal jacket for imparting maximum abrasion resistance to the completed cable. Braided armor is sometimes used in lieu of solid armor for improved flexibility.

AROMATIC HYDROCARBONS

Hydrocarbons derived from or characterized by presence of unsaturated resonant ring structures.

ARTIFICIAL AGEING

The accelerated testing of plastics specimens to determine their changes in properties. Carried out over a short period of time, such tests are indicative of what may be expected of a material under service conditions over extended periods. Typical investigations include those for dimensional stability; the effect of immersion in water, chemicals and solvents; light stability and resistance to fatigue.

ARTIFICIAL WEATHERING

Exposure to cyclic laboratory conditions involving changes in temperature, relative humidity and radiant energy, with or without direct water spray, in an attempt to produce changes in the material similar to those observed after long-term continuous outdoor exposure.

ASBESTOS

A gray, non-burning, non-conducting and chemical resistant amphibole occuring in long fibers or fibrous masses, sometimes used as a filler for reinforcement.

ATACTIC

A chain of molecules in which the position of the methyl groups is more or less random.

AUTOCLAVE

(1) Closed strong vessel for conducting chemical reactions under high pressure; (2) in low-pressure laminating, a round or cylindrical container in which heat and gas pressure can be appied to resin-impregnated paper or fabric positioned in layers over a mold.

AUTOCLAVE MOLDING

Modification of the pressure bag method for molding reinforced plastics. After lay-up, entire assembly is placed in steam autoclave at 50 to 100 psi. Additional pressure achieves higher reinforcement loadings and improved removal of air.

AUTOMATIC MOLD

A mold for injection or compression molding that repeatedly goes through the entire cycle, including ejection, without human assistance.

AVERAGE MOLECULAR WEIGHT (viscosity method)

The molecular weight of polymeric materials determined by the viscosity of the polymer in solution at a specific temperature. This gives an average molecular weight of the molecular chains in the polymer independent of specific chain length. Falls between weight average and number average molecular weight.

B

B-STAGE

An intermediate stage in the reaction of a thermosetting resin in which the material softens when heated and swells in contact with certain liquids but does not entirely fuse or dissolve. Resins in thermosetting molding compounds are usually in this stage. See also *A-Stage* and *C-Stage.*

BACKING PLATE

In injection molding, a plate used as a support for the cavity blocks, guide pins, bushings, etc.

BACK PRESSURE

The viscosity resistance of a material to continued flow when a mold is closing. In extrusion, the resistance to the forward flow of molten material.

BAFFLE

A device used to restrict or divert the passage of fluid through a pipe line or channel. In hydraulic systems the device, which often consists of a disc with a small central perforation, restricts the flow of hydraulic fluid in a high pressure line. A common location for the disc is in a joint in the line. When applied to molds, the term is indictative of a plug or similar device located in a steam or water channel in the mold and designed to divert and restrict the flow to a desired path.

BAG MOLDING

A method of applying pressure during bonding or molding, in which a flexible cover, usually in connection with a rigid die or mold, exerts pressure on the material being molded, through the application of air pressure or drawing of a vacuum.

BAKELITE

The proprietary name for phenolic and other plastics materials produced by Bakelite Limited, but often used indiscriminately to describe any phenolic molding material or molding. The name is derived from that of Dr. Leo Hendrik Baekeland (1863-1944), a Belgian who, through his work on the synthesis of phenolic resins and their commercial development in the early 1900's, is generally considered to be the "father" of the plastics industry.

BANBURY

An apparatus for compounding materials composed of a pair of contra-rotating rotors which masticate the materials to form a homogeneous blend. This is an internal type mixer which produces excellent mixing.

BENZENE RING

The basic structure of benzene, the most important aromatic chemical. It is an unsaturated, resonant 6-carbon ring having three double bonds. One or more of the 6 hydrogen atoms of benzene may be replaced by other atoms or groups.

BETA GAGE (or beta-ray gage)

A gage consisting of two facing elements, a B-ray-emitting source and a B-ray detector. When a sheet material is passed between the elements, some of the B-rays are absorbed, the percent absorbed being a measure of the areal density or the thickness of the sheet.

BINDER

In a reinforced plastic, the continuous phase which holds together the reinforcement.

BLANKING

The cutting of flat sheet stock to shape by striking it sharply with a punch while it is supported on a mating die. Punch presses are used. Also called *Die Cutting.*

BLEED

To give up color when in contact with water or a solvent; undesired movement of certain materials in a plastic (e.g. plasticizers in vinyl) to the surface of the finished article or into an adjacent material. Also called *Migration.*

BLISTER

A raised area on the surface of a molding caused by the pressure of gases inside it on its incompletely hardened surface.

BLOCK COPOLYMER

An essentially linear copolymer in which there are repeated sequences of polymeric segments of different chemical structure.

BLOCKING

An undesired adhesion between touching layers of a material, such as occurs under moderate pressure during storage or use.

BLOOM

A visible exudation or efflorescence on the surface of a material.

BLOW MOLDING

A method of fabrication in which a parison (hollow tube) is forced into the shape of the mold cavity by internal air pressure.

BLOW PRESSURE

The air pressure used to form a hollow part by blow molding.

BLOW RATE

The speed at which the air enters the parison during the blow molding cycle.

BLOWING AGENTS

See *Foaming Agents.*

BLOWUP RATIO

In blow molding, the ratio of the mold cavity diameter to the parison diameter. In blown tubing (film), the ratio of the final tube diameter (before gusseting, if any) to the original die diameter.

BLOWN TUBING

A thermoplastic film which is produced by extruding a tube, applying a slight internal pressure to the tube to expand it while still molten and subsequent cooling to set the tube. The tube is then flattened through guides and wound up flat on rolls. The size of blown tubing is determined by the flat width in inches as wound rather than by the diameter as in the case of rigid types of tubing.

BLUEING

A mold blemish in the form of a blue oxide film which occurs on the polished surface of a mold as a result of the use of abnormally high mold temperatures.

BLUNT THREAD START

A thread design where the start of the thread has been squared off for exact locating of the container in a printing or labeling machine. See *Lug.*

BOSS

Protuberance on a plastic part designed to add strength, to facilitate alignment during assembly, to provide for fastenings, etc.

BOSTON ROUND

A particular shape of container; cross section as well as shoulders are round.

BOTTOM BLOW

A specific type of blow molding machine which forms hollow articles by injecting the blowing air into the parison from the bottom of the mold.

BOTTOM PLATE

Part of the mold which contains the heel radius and the push-up.

BRANCHED

In molecular structure of polymers (as opposed to *Linear*), refers to side chains attached to the main chain. Side chains may be long or short.

BREAKDOWN VOLTAGE

The voltage required, under specific conditions, to cause the failure of an insulating material. See *Dielectric Strength.*

BREAKER PLATE

A perforated plate located at the rear end of an extruder head. It often supports the screens that prevent foreign particles from entering the die.

BREATHING

The opening and closing of a mold to allow gases to escape early in the molding cycle. Also called *Degassing.* When referring to plastic sheeting, "breathing" indicates permeability to air.

BUBBLER MOLD COOLING
(Injection Molding)

A method of cooling an injection mold in which a stream of cooling liquid flows continuously into a cooling cavity equipped with a coolant outlet normally positioned at the end opposite the inlet. Uniform cooling can be achieved in this manner.

BULK DENSITY

The mass per unit volume of a molding powder as determined in a reasonably large volume. The recommended test method is ASTM D1182-54.

BULK FACTOR

Ratio of the volume of loose molding powder to the volume of the same weight of resin after molding.

BURNING BEHAVIOR

The characteristics exhibited by the substance when it ignites and burns.

BURNING RATE

A term describing the tendency of plastics articles to burn at given temperatures. Certain plastics, such as those based on shellac, burn readily at comparatively low temperatures. Others will melt or disintegrate without actually burning, or will burn only if exposed to direct flame. These latter are often referred to as self-extinguishing.

BUSHING (Extrusion)

The outer ring of any type of a circular tubing or pipe die which forms the outer surface of the tube or pipe.

BUTADIENE

$CH_2:CH \cdot CH:CH_2$. A gas, insoluble in water but soluble in alcohol and ether, obtained from the cracking of petroleum, from coal tar benzene or from acetylene produced from coke and lime. It is widely used in the formation of copolymers with styrene, acrylonitrile, vinyl chloride and other monomeric substances, where it imparts flexibility to the subsequent moldings.

BUTADIENE STYRENE PLASTICS

A synthetic resin derived from the copolymerization of butadiene gas and styrene liquids.

BUTT-FUSION

A method of joining pipe, sheet, or other similar forms of a thermoplastic resin wherein the ends of the two pieces to be joined are heated to the molten state and then rapidly pressed together to form a homogeneous bond.

BUTTRESS THREAD

A type of threading in which the thread sides terminate abruptly in threading gradually tapering down to the neck finish. Designed to withstand maximum force in one direction only. Cross section of thread is triangular.

BUTYLENE PLASTICS

Plastics based on resins made by the polymerization of butene or copolymerization by butene with one or more unsaturated compounds, the butene being in greatest amount by weight.

C

C-STAGE

The final stage in the reactions of a thermosetting resin in which the material is relatively insoluble and infusible. Thermosetting resins in a fully cured plastic are in this stage. See *A-Stage* and *B-Stage*.

CABLE

A standard conductor; or a group of solid or standard conductors laid together but insulated from one another.

CALENDER

(*v.*) To prepare sheets of material by pressure between two or more counter-rotating rolls. (*n.*) — The machine performing this operation.

CAPROLACTAM

A cyclic amidetype compound, containing 6 carbon atoms. When the ring is opened, caprolactam is polymerizable into a nylon resin known as type-6 nylon or polycaprolactam.

CARBON BLACK

A black pigment produced by the incomplete burning of natural gas or oil. It is widely used as a filler, particularly in the rubber industry. Because it possesses useful ultraviolet protective properties, it is also much used in polyethylene compounds intended for such applications as cold water piping and black agricultural sheet.

CASEIN

A protein material precipitated from skimmed milk by the action of either rennet or dilute acid. Rennet casein finds its main application in the manufacture of plastics. Acid casein is a raw material used in a number of industries including the manufacture of adhesives.

CAST

(1) To form a "plastic" object by pouring a fluid monomer-polymer solution into an open mold where it finishes polymerizing. (2) Forming plastic film and sheet by pouring the liquid resin onto a moving belt or by precipitation in a chemical bath.

CAST FILM

A film made by depositing a layer of plastic, either molten, in solution, or in a dispersion, onto a surface, solidifying and removing the film from the surface.

CASTING (n.)

The finished product of a casting operation; should not be used for molding, q.v.

CASTING AREA

The moldable area of a thermoplastic in square inches for a given thickness and under a given set of injection molding conditions. Casting area is a measure of flow under actual molding conditions where flow is unrestricted by cavity boundaries.

CATALYST

A substance which markedly speeds up the cure of a compound when added in minor quantity as compared to the amounts of primary reactants. See *Hardener, Inhibitor, Promoter.*

CAVITY

Depression in a mold made by casting, machining, hobbing, or a combination of these methods; depending on number of such depressions, molds are designated as *Single-Cavity* or *Multi-Cavity.*

CELL

A small particle or completely enclosed cavity.

CELL (closed)

A cell totally enclosed by its walls and hence not interconnecting with other cells. (See *Cell* and *Cell, open).*

CELL (open)

A cell not totally enclosed by its walls and hence interconnecting with other cells. (See *Cell* and *Cell, closed).*

CELLULAR PLASTICS

Plastics containing numerous small cavities (cells), interconnecting or not distributed throughout the mass.

CELLULOID

A thermoplastic material made by the intimate blending of cellulose nitrate, q.v., with camphor. Alcohol is normally employed as a volatile solvent to assist plasticization, and is subsequently removed.

CELLULOSE

A natural high polymeric carbohydrate found in most plants; the main constituent of dried woods, jute, flax, hemp, ramie, etc. Cotton is almost pure cellulose.

CELLULOSE ACETATE

An acetic acid ester of cellulose. It is obtained by the action, under rigidly controlled conditions, of acetic acid and acetic anhydride on purified cellulose usually obtained from cotton linters. All three available hydroxyl groups in each glucose unit of the cellulose can be acetylated but in the material normally used for plastics it is usual to acetylate fully and then to lower the acetyl value (expressed as acetic acid) to 52-56% by partial hydrolysis. When compounded with suitable plasticizers it gives a tough thermoplastic material.

CELLULOSE ACETATE BUTYRATE

An ester of cellulose made by the action of a mixture of acetic and butyric acids and their anhydrides on purified cellulose. It is used in the manufacture of plastics which are similar in general properties to cellulose acetate but are tougher and have better moisture resistance and dimensional stability.

CELLULOSE ESTER

A derivative of cellulose in which the free hydroxyl groups attached to the cellulose chain have been replaced wholly or in part by acetic groups, e.g., nitrate, acetate, or stearate groups. Esterification is effected by the use of a mixture of an acid with its anhydride in the presence of a catalyst, such as sulfuric acid. Mixed esters of cellulose, e.g., cellulose acetate butyrate, are prepared by the use of mixed acids and mixed anhydrides. Esters and mixed esters, a wide range of which is known, differ in their compatibility with plasticizers, in molding properties, and in physical characteristics. These esters and mixed esters are used in the manufacture of thermoplastic molding compositions.

CELLULOSE NITRATE
(Nitrocellulose)

A nitric acid ester of cellulose manufactured by the action of a mixture of sulfuric acid and nitric acid on cellulose, such as purified cotton linters. The type of cellulose nitrate used for celluloid manufacture usually contains 10.8-11.1% of nitrogen. The latter figure is the nitrogen content of the dinitrate.

CELLULOSE PROPIONATE

An ester of cellulose made by the action of propionic acid and its anhydride on purified cellulose. It is used as the basis of a thermoplastic molding material.

CELLULOSE TRIACETATE

A cellulosic material made by reacting purified cellulose with acetic anhydride in the presence of a catalyst. It is used in the form of film and fibers. Films and sheet are cast from clear solutions on to "drums" with highly polished surfaces. The film, which is of excellent clarity, has high tensile strength, and good heat resistance and dimensional stability. Applications include book jackets, magnetic recording tapes, and various types of packaging. Cellulose triacetate sheet has somewhat similar properties to those of the film and is used to make such articles as safety goggles, map wallets and transparent covers of many kinds.

CENTER GATED MOLD

An injection mold wherein the cavity is filled with resin through an orifice interconnecting the nozzle and the center of the cavity area. Normally, this orifice is located at the bottom of the cavity when forming items such as containers, tumblers, bowls, etc.

CENTRIFUGAL CASTING

A method of forming thermoplastic resins in which the granular resin is placed in a rotatable container, heated to a molten condition by the transfer of heat through the walls of the container, and rotated so that the centrifugal force induced will force the molten resin to conform to the configuration of the interior surface of the container. Used to fabricate large diameter pipes and similar cylindrical items.

CHAIN LENGTH

See *Degree Of Polymerization*.

CHALKING

A powdery residue on the surface of a material often resulting from degradation.

CHARGE

The measurement or weight of material used to load a mold at one time or during one cycle.

CHASE

An enclosure of any shape, used to: (a) shrink-fit parts of a mold cavity in place; (b) prevent spreading or distortion in hobbing; (c) enclose an assembly of two or more parts of a split cavity block.

CHEMICALLY FOAMED POLYMERIC MATERIAL

A cellular material in which the cells are formed by gases generated from thermal decomposition or other chemical reaction.

CHILL ROLL

A cored roll, usually temperature controlled with circulating water, which cools the web before winding. For chill roll (cast) film, the surface of the roll is highly polished. In extrusion coating, either a polished or a matte surface may be used depending on the surface desired on the finished coating.

CHILL ROLL EXTRUSION
(or cast film extrusion)

The extruded film is cooled while being drawn around two or more highly polished chill rolls cored for water cooling for exact temperature control.

CHLORINATED POLYETHER

The polymer is obtained from pentaerythritol by preparing a chlorinated oxetane and polymerizing it to a polyether by means of opening the ring structure.

CHOKED NECK

Narrowed or constricted opening in the neck of a container.

CHROMIUM PLATING

An electrolytic process that deposits a hard film of chromium metal onto working surfaces of other metals where resistance to corrosion, abrasion, and/or erosion is needed.

CHLORINATED POLYVINYL CHLORIDE PLASTICS

Plastics based on chlorinated polyvinyl chloride in which the chlorinated polyvinyl chloride is in the greatest amount by weight.

CHLOROFLUOROCARBON PLASTICS

Plastics based on polymers made with monomers composed of chlorine, fluorine, and carbon only.

CIL (Flow Test)

A method of determining the rheology of flow properties of thermoplastic resins developed by Canadian Industries Limited. In this test, the amount of the molten resin which is forced through a specified size orifice per unit time when a specified, variable force is applied gives a relative indication of the flow properties of various resins.

CIRCUIT

In filament winding, the winding produced by a single revolution of mandrel or form.

CLAMPING PLATE

A plate fitted to a mold and used to fasten the mold to a molding machine.

CLAMPING PRESSURE

In injection molding and in transfer molding, the pressure which is applied to the mold to keep it closed, in opposition to the fluid pressure of the compressed molding material.

CLARIFIER

An additive that increases the transparency of a material.

CLEARANCE

A controlled distance by which one part of an object is kept separated from another part.

COALESCE

To combine into one body or to grow together.

COATING

See specific type of coating such as *Curtain, Extrusion, Kissroll spray.*

COATING WEIGHT

The weight of coating per unit area. In the United States usually "per ream," i.e., 500 sheets 24" x 36" (3000 sq. ft.), but sometimes 1000 sq. ft.

COEFFICIENT OF EXPANSION

The fractional change in length (sometimes volume, specified) of a material for a unit change in temperature. Values for plastics range from 0.01 to 0.2 mils/in., $^{\circ}$C.

COEFFICIENT OF FRICTION

See *Friction Coefficient.*

COLD FLOW

See *Creep.*

COLD MOLDING

A procedure in which a composition is shaped at room temperature and cured by subsequent baking.

COLD SLUG

The first material to enter an injection mold; so called because in passing through sprue orifice it is cooled below the effective molding temperature.

COLD SLUG WELL

Space provided directly opposite the sprue opening in an injection mold to trap the cold slug.

COLD STRETCH

Pulling operation, usually on extruded filaments, to improve tensile properties.

COLLAPSE

Contraction of the walls of a container, e.g., upon cooling, leading to a permanent indentation.

COMPOUND

An intimate admixture of (a) polymer(s) with all the materials necessary for the finished product.

COMPRESSION MOLD

A mold which is open when the material is introduced and which shapes the material by heat and by the pressure of closing.

COMPRESSION MOLDING

A technique of thermoset molding in which the molding compound (generally preheated) is placed in the open mold cavity, mold is closed, and heat and pressure (in the form of a downward moving ram) are applied until the material has cured.

COMPRESSION RATIO

In an extruder screw, the ratio of volume available in the first flight at the hopper to the last flight at the end of the screw.

COMPRESSIVE STRENGTH

Crushing load at the failure of a specimen divided by the original sectional area of the specimen.

CONCENTRICITY

For a container, the shape in which various cross sections have a common center.

CONDENSATION

A chemical reaction in which two or more molecules combine with the separation of water or some other simple substance. If a polymer is formed, the condensation process is called *Polycondensation.* See also *Polymerization.*

CONDENSATION RESIN

A resin formed by polycondensation, e.g., the alkyd, phenolaldehyde, and urea formaldehyde resins.

CONDITIONING

The subjection of a material to a stipulated treatment so that it will respond in a uniform way to subsequent testing or processing. The term is frequently used to refer to the treatment given to specimens before testing.

CONDUCTOR

A wire, or combination of wires not insulated from each other, suitable for carrying electricity.

CONTACT PRESSURE RESINS

Liquid resins which thicken or resinify on heating and, when used for bonding laminates, require little or no pressure.

CONTINUOUS THREAD

A spiral, protruding finish on the neck of a container to hold a screw-type closure.

CONVERGENT DIE

A die in which the internal channels leading to the orifice are converging (only applicable to dies for hollow bodies).

CONVEYOR

A mechanical device to transport material from one point to another, often continuously.

COOLING CHANNELS

Channels or passageways located within the body of a mold through which a cooling medium can be circulated to control temperature on the mold surface.

COOLING FIXTURE

Block of metal or wood holding the shape of a molded piece which is used to maintain the proper shape or dimensional accuracy of a molding after it is removed from the mold until it is cool enough to retain its shape without further appreciable distortion. Also known as *Shrink Fixture*.

COPOLYMER

See *Polymer*.

CORE

(1) The central member of a sandwich construction (can be honeycomb material, foamed plastic, or solid sheet) to which the faces of the sandwich are attached; the central member of a plywood assembly. (2) A channel in a mold for circulation of heat-transfer media. (3) Part of a complex mold

that molds undercut parts. Cores are usually withdrawn to one side before the main sections of the mold open. Also called *Core Pin*.

CORE DRILL

A device for making cooling channels in a mold.

CORONA RESISTANCE

A current passing through a conductor induces a surrounding electrostatic field. When voids exist in the insulation near the conductor, the high voltage electrostatic field may ionize and rapidly accelerate some of the air molecules in the void. These ions can then collide with the other molecules, ionizing them, and thereby "eating" a hole in the insulation. Resistance to this process is called corona resistance.

CORROSION RESISTANCE

The ability to withstand the effect of oxidation.

COVER

In wire coating, a coating whose primary purpose is to "weatherproof" or to prevent casual grounding (such as contact with a wet tree branch), or to otherwise protect a conductor.

CRATER

A small, shallow surface imperfection.

CRAZING

Fine cracks which may extend in a network on or under the surface or through a layer of a plastic material.

CREEP

The dimensional change with time of a material under load, following the initial instantaneous elastic deformation. Creep at room temperature is sometimes called *Cold Flow*.

CROSSHEAD (Extrusion)

A device generally employed in wire coating which is attached to the discharge end of the extruder cylinder, designed to facilitate extruding material at an angle. Normally, this is a 90 degree angle to the longitudinal axis of the screw.

CROSS LAMINATE

A laminate in which some of the layers of material are oriented approximately at right angles to the remaining layers with respect to the grain or strongest direction in tension.

CROSS-LINKING

Applied to polymer molecules, the setting-up of chemical links between the molecular chains. When extensive, as in most thermosetting resins, cross-linking makes one infusible super-molecule of all the chains.

CRYSTALLINITY

A state of molecular structure in some resins which denotes uniformity and compactness of the molecular chains forming the polymer. Normally can be attributed to the formation of solid crystals having a definite geometric form.

CULL

Material remaining in a transfer chamber after mold has been filled. Unless there is a slight excess in the charge, the operator cannot be sure cavity is filled. Charge is generally regulated to control thickness of cull.

CURE

To change the properties of a polymeric system into a more stable, usable condition by the use of heat, radiation, or reaction with chemical additives.

NOTE — Cure may be accomplished, for example, by removal of solvent or cross-linking.

CURE CYCLE

The schedule of time periods at specified conditions to which a reacting thermosetting material is subjected to reach a specified property level.

CURING TEMPERATURE

Temperature at which a cast, molded, or extruded product, a resin-impregnated reinforcing material, an adhesive, etc., is subjected to curing.

CURE TIME

The period of time that a reacting thermosetting material is exposed to specific conditions to reach a specified property level.

CURLING

A condition in which the parison curls upwards and outwards, sticking to the outer face of the die ring. Balance of temperatures between die and mandrel will normally relieve this problem.

CURTAIN COATING

A method of coating which may be employed with low viscosity resins or solutions, suspensions, or emulsions of resins in which the substrate to be coated is passed through and perpendicular to a freely falling liquid "curtain" (or "waterfall"). The flow rate of the falling liquid and the linear speed of the substrate passing through the curtain are coordinated in accordance with the thickness of coating desired.

CURVATURE

A condition in which the parison is not straight, but somewhat bending and shifting to one side, leading to a deviation from the vertical direction of extrusion. Centering of ring and mandrel can often relieve this defect.

CUT-OFF

The line where the two halves of a compression mold come together; also called *Flash Groove* or *Pinch-off.*

CYCLE

The complete, repeating sequence of operations in a process or part of a process. In molding, the cycle time is the period, or elaspsed time, between a certain point in one cycle and the same point in the next.

CYLINDRICAL

Refers to the shape of a container which has a circular cross section parallel to the minor axis and a rectangular cross section parallel to the major axis.

D

DASH-POT

A device used in hydraulic systems for damping down vibration. It consists of a piston attached to the part to be damped and fitted into a vessel containing fluid or air. It absorbs shocks by reducing the rate of change in the momentum of moving parts of machinery.

DAYLIGHT OPENING

Clearance between two platens of a press in the open position.

DEBOSSED

An indent or cut in design or lettering of a surface.

DECKLE ROD

A small rod, or similar device, inserted at each end of the extrusion coating die which is used to adjust the length of the die opening.

DECOMPOSITION PRODUCT

The constituent elements or simpler compounds formed when a substance decays or decomposes.

DECORATIVE SHEET

A laminated plastics sheet used for decorative purposes in which the color and/or surface pattern is an integral part of the sheet.

DEFLASHING

Covers the range of finishing techniques used to remove the flash (excess, unwanted material) on a plastic molding.

DEGASSING

See *Breathing*.

DEGRADATION

A deleterious change in the chemical structure of a plastic.

DEGREE OF POLYMERIZATION (DP)

The number of structural units or mers in the "average" polymer molecule in a particular sample. In most plastics the DP must reach several thousand if worthwhile physical properties are to be had.

DELAMINATION

The separation of the layers in a laminate caused by the failure of the adhesive.

DELIQUESCENT

Capable of attracting moisture from the air.

DENIER

The weight (in grams) of 9000 meters of synthetic fiber in the form of continuous filament.

DENSITY

Weight per unit volume of a substance, expressed in grams per cubic centimeter, pounds per cubic foot, etc.

DESICCANT

Substance which can be used for drying purposes because of its affinity for water.

DESIGN BASIS

Term used for long term strength of plastic pipe determined in accordance with ASTM D2837.

DESTATICIZATION

Treating plastics materials to minimize their accumulation of static electricity and, consequently, the amount of dust picked up by the plastics because of such charges.

DETERGENTS

Substances with a high surface activity. Similar in general to soaps; made synthetically to a large extent.

DIE BLADES

Deformable member(s) attached to a die body which determine the slot opening and which are adjusted to produce uniform thickness across the film or sheet produced.

DIE CUTTING

(1) Blanking q.v., (2) Cutting shapes from sheet stock by striking it sharply with a shaped inife edge known as a XXsteel-rule die.'' Clicking and Dinking are other names for die cutting of this kind.

DIE GAP

The distance between the metal faces forming the die opening.

DIELECTRIC

Insulating material. In radio-frequency preheating, dielectric may refer specifically to the material which is being heated.

DIELECTRIC CONSTANT

Normally the relative dielectric constant; for practical purposes, the ratio of the capacitance of an assembly of two electrodes separated solely by a plastics insulating material to its capacitance when the electrodes are separated by air (ASTM D150-59T).

DIELECTRIC HEATING
(Electronic heating)

The plastic to be heated forms the dielectric of a condenser to which is applied a high-frequency (20 to 80 mc.) voltage. Dielectric loss in the material is the basis. Process used for sealing vinyl films and preheating thermoset molding compounds.

DIELECTRIC STRENGTH

The electric voltage gradient at which an insulating material is broken down or "arced through," in volts per mil of thickness.

DIE LINES

Vertical marks on the parison caused by damage of die parts or contamination.

DIE SWELL RATIO

The ratio of the outer parison diameter (or parison thickness) to the outer diameter of the die (or die gap). Die swell ratio is influenced by polymer type, head construction, land length, extrusion speed, and temperature.

DIMENSIONAL STABILITY

Ability of a plastic part to retain the precise shape in which it was molded, fabricated, or cast.

DIMER

A substance (comprising molecules) formed from two molecules of a monomer.

DIP COATING

Applying a plastic coating by dipping the article to be coated into a tank of melted resin or plastisol, then chilling the adhering melt.

DISCOLORATION

Any change from the original color, often caused by overheating, light exposure, irradiation, or chemical attack.

DISPERSION

Finely divided particles of a material in suspension in another substance.

DISSIPATION FACTOR
See *Power Factor*.

DIVERGENT DIE
A die in which the internal channels leading to the orifice are diverging (applicable only to dies for hollow bodies).

DOME
In reinforced plastics, an end of a filament wound cylindrical container.

DOUBLE-SHOT MOLDING
A means of turning out two-color parts in thermoplastics materials by successive molding operations.

DRAFT
The degree of taper of a side wall or the angle of clearance designed to facilitate removal of parts from a mold.

DRAPE ASSIST FRAME
In sheet thermoforming, a frame (made up of anything from thin wires to thick bars) shaped to the peripheries of the depressed areas of the mold and suspended above the sheet to be formed. During forming, the assist frame drops down, drawing the sheet tightly into the mold and thereby preventing webbing between high areas of the mold and permitting closer spacing in multiple molds.

DRAPE FORMING
Method of forming thermoplastic sheet in which the sheet is clamped into a movable frame, heated, and draped over high points of a male mold. Vacuum is then pulled to complete the forming operation.

DRAW DOWN RATIO
The ratio of the thickness of the die opening to the final thickness of the product.

DRAWING
The process of stretching a thermoplastic sheet or rod to reduce its cross-sectional area.

DRY-BLEND
A free-flowing dry compound prepared without fluxing or addition of solvent also called powder blend.

DRY COLORING
Method commonly used by fabricators for coloring plastics by tumble blending uncolored particles of the plastic material with selected dyes and pigments.

DRY STRENGTH
The strength of an adhesive joint determined immediately after drying under specified conditions or after a period of conditioning in the standard laboratory atmosphere. See *Wet Strength*.

DUCTILITY
The extent to which a solid material can be drawn into a thinner cross section.

DWELL
A pause in the application of pressure to a mold, made just before the mold is completely closed, to allow the escape of gas from the molding material.

DYES
Synthetic or natural organic chemicals that are soluble in most common solvents. Characterized by good transparency, high tinctorial strength, and low specific gravity.

E

EFFECTIVE THREAD TURNS
The number of full $360°$ turns on a threaded closure that are actually in contact with the neck thread.

EJECTOR PIN (on sleeve)

A pin or thin plate that is driven into a mold cavity from the rear as the mold opens, forcing out the finished piece. Also *Knockout Pin*.

EJECTOR RETURN PINS

Projections that push the ejector assembly back as the mold closes; also called *Surface Pins* and *Return Pins*.

EJECTOR ROD

Bar that actuates the ejector assembly when mold is opened.

ELASTIC DEFORMATION

The part of the deformation of an object under load which is recoverable when the load is removed.

ELASTOMER

A material which at room temperature stretches under low stress to at least twice its length and snaps back to the original length upon release of stress. See also *Rubber*.

ELECTROFORMED MOLDS

A mold made by electroplating metal on the reverse pattern on the cavity. Molten steel may be then sprayed on the back of the mold to increase its strength.

ELECTRONIC TREATING

A method of oxidizing a film of polyethylene to render it printable by passing the film between the electrodes and subjecting it to a high voltage corona discharge.

ELECTROPLATING

The deposition of a layer of metal on a base of metal or conducting surface by electrolysis.

ELONGATION

The fractional increase in length of a material stressed in tension.

EMBOSSING

Techniques used to create depressions of a specific pattern in plastics film and sheeting.

EMULSION

A suspension of fine droplets of one liquid in another.

ENCAPSULATING

Enclosing an article (usually an electronic component or the like) in a closed envelope of plastic, by immersing the object in a casting resin and allowing the resin to polymerize or, if hot, to cool.

ENDOTHERMIC REACTION

A chemical reaction in which heat is absorbed.

ENGRAVED-ROLL (or Gravure) COATING

The amount of coating applied to the web is metered by the depth of the over-all engraved pattern in a print roll. This process is frequently modified by interposing a resilient offset roll between the engraved roll and the web.

ENTRANCE ANGLE

Maximum angle at which the molten material enters the land area of the die, measured from the center line of the mandrel.

ENVIRONMENTAL STRESS CRACKING (ESG)

The susceptibility of a thermoplastic article to crack or craze formation under the influence of certain chemicals and stress.

EPOXY RESINS

Based on ethylene oxide, its derivatives or homologs, epoxy resins form straight-chain thermoplastics and thermosetting resins, e.g., by the condensation of bisphenol and epichlorohydrin.

ESC or ESCR
See *Environmental Stress Cracking.*

ESTER
The reaction product of an alcohol and an acid.

ETHYLENE PLASTICS
Plastics based on polymers of ethylene or copolymers of ethylene with other monomers, the ethylene being in greatest amount by mass.

ETHYLENE-VINYL ACETATE
Copolymers from these two monomers form a new class of plastic materials. They retain many of the properties of polyethylene, but have considerably increased flexibility for their density — elongation and impact resistance are also increased.

EXOTHERM
(1) The temperature/time curve of a chemical reaction giving off heat, particularly the polymerization of casting resins. (2) The amount of heat given off. The term has not been standardized with respect to sample size, ambient temperature, degree of mixing, etc.

EXOTHERMIC REACTION
A chemical reaction in which heat is evolved.

EXPANDED PLASTICS
See *Open-cell Foamed Plastics.*

EXTENDER
A substance generally having some adhesive action, added to a plastic composition to reduce the amount of the primary resin required per unit area.

EXTRUDATE
The product or material delivered by an extruder, such as film, pipe, the coating on wire, etc.

EXTRUSION
The compacting of a plastic material and the forcing of it through an orifice in more or less continous fashion.

EXTRUSION COATING
The resin is coated on a substrate by extruding a thin film of molten resin and pressing it onto or into the substrates, or both, without the use of an adhesive.

F

FABRICATE
To work a material into a finished form by machining, forming, or other operation or to make flexible film or sheeting into end-products by sewing, cutting, sealing, or other operation.

FADEOMETER
An apparatus for determining the resistance of resins and other materials to fading. This apparatus accelerates the fading by subjecting the article to high intensity ultraviolet rays of approximately the same wave length as those found in sunlight.

FALSE NECK
A neck construction which is additional to the neck finish of a container and which is only intended to faciltate the blow molding operation. Afterwards the false neck part is removed from the container.

FAMILY MOLD (Injection)
A multi-cavity mold wherein each of the cavities forms one of the component parts of the assembled finished object.

FATTY ACID
An organic acid obtained by the hydrolysis (saponification) of natural fats and oils, e.g., stearic and

palmitic acids. These acids are monobasic, may or may not have some double bonds, contain 16 or more C atoms.

FAULT

An electrical short circuit or leakage path to ground or from phase to phase inadvertently created.

FEMALE

In molding practice, the indented half of a mold designed to receive the male half.

FIBER

This term usually refers to relatively short lengths of very small cross-sections of various materials. Fibers can be made by chopping filaments (converting). Staple fibers may be ½ to a few inches in length and usually 1 to 5 denier (½ to 1 mils in diameter in Marlex polyethylene).

FIBER SHOW

Strands or bundles of fibers that are not covered by resin and that are at or above the surface of a reinforced plastic.

FILAMENT

A variety of fiber characterized by extreme length, which permits its use in yarn with little or no twist and usually without the spinning operation required for fibers.

FILAMENT WINDING

Roving or single strands of glass, metal, or other reinforcement are wound in a predetermined pattern onto a suitable mandrel. The pattern is so designed as to give maximum strength in the directions required. The strands can either be run from a creel through a resin bath before winding or pre-impregnated materials can be used. When the right number of layers have been applied, the wound mandrel is cured at room temperatures or in an oven.

FILL-AND-WIPE

Parts are molded with depressed designs; after application of paint, surplus is wiped off, leaving paint remaining only in depressed areas.

FILL POINT

The level to which a container must be filled to furnish a designated quantity of the content.

FILLER

A cheap, inert substance added to a plastic to make it less costly. Fillers may also improve physical properties, particularly hardness, stiffness, and impact strength. The particles are usually small, in contrast to those of reinforcements q.v., but there is some overlap between the functions of the two.

FILLET

A rounded filling of the internal angle between two surfaces of a plastic molding.

FILM

An optional term for sheeting having a nominal thickness not greater than 0.010 inch.

FIN

The web of material remaining in holes or openings in a molded part which must be removed in finishing.

FINES

Very small particles (usually under 200 mesh) accompanying larger grains, usually of molding powder.

FINISH

The plastic forming the opening of a container shaped to accommodate a specific closure. Also, the ultimate surface structure of an article.

FINISH INSERT

A removable part of a blow mold to form a specific neck finish of a plastic bottle. Sometimes called *Neck Insert*.

FISH EYE

A fault in transparent or translucent plastics materials, such as film or sheet, appearing as a small globular mass and caused by incomplete blending of the mass with surrounding material.

FITMENT

A device used as a part of a closure assembly to accomplish certain purposes such as a dropper sprinker, powder shaker, etc.

FLAKE

Used to denote the dry, unplasticized base of cellulosic plastics.

FLAME RETARDANT RESIN

A resin which is compounded with certain chemicals to reduce or eliminate its tendency to burn. For polyethylene and similar resins, chemicals such as antimony trioxide and chlorinated paraffins are useful.

FLAME SPRAYING

Method of applying a plastic coating in which finely powdered fragments of the plastic, together with suitable fluxes, are projected through a cone of flame onto a surface.

FLAME TREATING

A method of rendering inert thermoplastic objects receptive to inks, lacquers, paints, adhesives, etc. in which the object is bathed in an open flame to promote oxidation of the surface of the article.

FLAMMABILITY

Measure of the extent to which a material will support combustion.

FLASH

Extra plastic attached to a molding along the parting line; it must

be removed before the part can be considered finished.

FLASH GATE

A long, shallow rectangular gate.

FLASH LINE

A raised line appearing on the surface of a molding and formed at the junction of mold faces.

FLASH MOLD

A mold designed to permit excess molding material to escape during closing.

FLASH POINT

The lowest temperature at which a combustible liquid will give off a flammable vapor that will burn momentarily.

FLEXIBLE MOLDS

Molds made of rubber or elastomeric plastics used for casting plastics. They can be stretched to remove cured pieces with undercuts.

FLEXIBILIZER

An additive that makes a resin or rubber more flexible, i.e., less stiff. Also a *Plasticizer*.

FLEXOGRAPHIC PRINTING

A rubber roll, partially immersed in an ink fountain, transfers the ink to a fine, screen-lined steel roller which, in turn, deposits a thin layer of ink on the printing plate.

FLEXURAL MODULUS

A measure of the strain imposed in the outermost fibers of a bent specimen.

FLEXURAL STRENGTH

The strength of a material in blending, expressed as the tensile stress of the outermost fibers of a bent test sample at the instant of failure. With plastics, this value is usually higher than the straight tensile strength.

FLOATING PLATEN

A platen located between the main head and the press table in a multi-daylight press and capable of being moved independently of them.

FLOCK

Short fibers of cotton, etc., used as fillers, q.v., for molding materials.

FLOCKING

A method of coating by spraying finely dispersed powders or fibers.

FLOW

A qualitative description of the fluidity of a plastic material during the process of molding.

FLOW LINE (weld line)

A mark on a molded piece made by the meeting of two flow fronts during molding.

FLOW MARKS

Wavy surface appearance of an object molded from thermoplastic resins caused by improper flow of the resin into the mold.

FLUIDIZED BED COATING

A method of applying a coating of a thermoplastic resin to an article in which the heated article is immersed in a dense-phase fluidized bed of powdered resin and thereafter heated in an oven to provide a smooth, pin-hole-free coating.

FLUORESCENT PIGMENTS

By absorbing unwanted wavelengths of light and converting them into light of desired wavelengths, these colors seem to possess an actual glow of their own.

FLUORINATED ETHYLENE PROPYLENE (FEP)

A member of the fluorocarbons, q.v., family of plastics, it is a copolymer of tetrafluoroethylene and hexafluoropropylene, possessing most of the properties of polytetrafluoroethylene (PTFE), q.v., and also having a melt viscosity low enough to permit conventional thermoplastic processing. Available in pellet form for molding and extrusion, and as dispersions for spray or dip-coating processes.

FLUORINE (F)

The most reactive non-metallic element. A pale yellow gas which is both corrosive and poisonous, it reacts vigorously with most oxidizable substances at room temperature, and forms fluorides. It is used in the production of metallic and other fluorides, some of which are used to introduce fluorine into organic compounds, i.e., the fluorocarbons, q.v.

FLUOROCARBON PLASTICS

Plastics based on polymers made with monomers composed of fluorine and carbon only.

FLUOROCARBONS

The family of plastics including polytetrafluoroethylene (PTFE); polychlorotrifluoroethylene (PCTFE); polyvinylidene and fluorinated ethylene propylene (FEP), q.v. They are characterized by properties including good thermal and chemical resistance and non-adhesiveness, and possess a low dissipation factor and low dielectric constant. Depending upon which of the fluorocarbons is used, they are available as molding materials, extrusion materials, dispersions, film or tape.

FLUOROPLASTICS

Plastics based on polymer with monomers containing one or more atoms of fluorine or copolymers of such monomers with other monomers, the fluorine containing monomer(s) being in greatest amount by mass.

FLUX

(1) An additive to a plastics composition during processing to improve its flow. For example, coumarone-indene resins are used as a flux during the milling of vinyl polymers. (2) Indicating a state of fluidity.

FOAMING AGENTS

Chemicals added to plastics and rubbers that generate inert gases on heating, causing the resin to assume a cellular structure.

FOAM-IN-PLACE

Refers to the deposition of foams which requires that the foaming machine be brought to the work which is "in place" as opposed to bringing the work to the foaming machine.

FOIL DECORATING

Molding paper, textile, or plastic foils printed with compatible inks directly into a plastic part so that the foil is visible below the surface of the part as integral decoration.

FORCE PLATE

The plate that carries the plunger or force plug of a mold and guide pins or bushings. Since it is usually drilled for steam or water lines, it is also called the *Steam Plate.*

FORCE PLUG

The portion of a mold that enters the cavity block and exerts pressure on the molding compound, designated as *Top Force* or *Bottom Force* by position in the assembly; also called *Plunger* or *Piston.*

FORMALDEHYDE (HCHO)

A colorless gas (usually employed as a solution in water) which possesses a suffocating, pungent odor. It is derived from the oxidation of methanol or low-boiling petroleum gases such as methane, ethane, propane and butane. It is widely used in the production of phenol formaldehyde (phenolic), urea formaldehyde (urea), and melamine formaldehyde (melamine) resins.

FREE SINTERING

See *Sintering.*

FRICTION CALENDERING

A process whereby an elastomeric compound is forced into the interstices of woven or cord fabrics while passing through the rolls of calender.

FRICTION COEFFICIENT

A number expressing the amount of frictional effect.

FRICTION WELDING

A method of welding thermoplastics materials whereby the heat necessary to soften the components is provided by friction.

FROST LINE

In the extrusion of polyethylene lay-flat film, a ring-shaped zone located at the point where the film reaches its final diameter. This zone is characterized by a "frosty" appearance to the film caused by the film temperature falling below the softening range of the resin.

FROTHING

Technique for applying urethane foam in which blowing agents or tiny air bubbles are introduced under pressure into the liquid mixture of foam ingredients.

FURAN RESINS

Dark colored, thermosetting resins available primarily as liquids ranging from low-viscosity polymers to thick, heavy syrups.

FURFURAL RESIN

A dark-colored synthetic resin of the thermosetting variety obtained by the condensation of furfural with phenol or its homologues. It is used in the manufacture of molding materials, adhesives and impregnating varnishes. Properties include high resistance to acids and alkalis.

FUSE

In plastisol molding, to heat the plastisol to the temperature at which it becomes a single homogeneous phase. In this sense, *Cure* is the same as *Fuse.*

G

GATE

In injection and transfer molding, the orifice through which the melt enters the cavity. Sometimes the gate has the same cross-section as the runner leading to it; often, it is severly restricted.

GEL

(*n.*) In polyethylene, a small amorphous resin particle which differs from its surroundings by being of higher molecular weight and/or crosslinked, so that its processing characteristics differ from the surrounding resin to such a degree that it is not easily dispersed in the surrounding resin. A gel is readily discernible in thin films.

GEL COAT

A thin, outer layer of resin, sometimes containing pigment, applied to a reinforced plastics molding as a cosmetic.

GEL POINT

The stage at which a liquid begins to exhibit pseudo-elastic properties.

GEL TIME

The time required for a liquid material to form a gel under specified conditions of temperature as measured by a specific test.

GLASS THREAD

A type of threading in which the thread sides gradually taper down to the neck finish. Cross section of threads are semi-circular.

GLASS TRANSITION

The reversible change in an amorphous polymer or in amorphous regions of a partially crystalline polymer from (or to) a viscous or rubbery condition to (or from) a hard and relatively brittle one.

NOTE — The glass transition generally occurs over a relatively narrow temperature region and is similar to the solidification of a liquid to a glassy state: it is not a phase transition. Not only do hardness and brittleness undergo rapid changes in this temperature region but other properties, such as thermal expansibility and specific heat also change rapidly. This phenomenon has been called second order transition, rubber transition and rubbery transition. The word transformation has also been used instead of transition. Where more than one amorphous transition occurs in a polymer, the one associated with segmental motions of the polymer backbone chain or accompanied by the largest change in properties is usually considered to be the glass transition.

GLASS TRANSITION TEMPERATURE (Tg)

The approximate midpoint of the temperature range over which the glass transition takes place.

GLITTER (or Flitter or Spangles)

A group of special decorative materials consisting of flakes large enough so that each separate flake produces a plainly visible sparkle or reflection. They are incorporated directly into the plastic during compounding.

GLOSS

The shine or lustre of the surface of a material.

GRADUATED

Molded in scale to indicate content level in container.

GRAFT COPOLYMERS

A chain of one type of polymer to which side chains of a different type are attached or grafted (i.e., polymerizing butadiene and styrene monomer at the same time).

GRANULAR STRUCTURE

Nonuniform appearance of finished plastic material due to retention of, or incomplete fusion of, particles of composition, either within the mass or on the surface.

GRAVURE PRINTING

Depositing ink on plastic film or sheeting or product from depressions of a specific depth, pattern, and spacing, which have been either mechanically or chemically engraved into a printing cylinder.

GRIT BLASTED

A surface treatment of a mold in which steel grit or sand materials are blown to the walls of the cavity to produce a roughened surface. Air escape from mold is improved and special appearance of molded article is often obtained by this method.

GUIDE PINS

Devices that maintain proper alignment of force plug and cavity as mold closes.

GUM

An amorphous substance or mixture which, at ordinary temperatures, is either a very viscous liquid or a solid which softens gradually on heating, and which either swells in water or is soluble in it. Natural gums, obtained from the cell walls of plants, are carbohydrates or carbohydrate derivatives of intermediate molecular weight.

GUSSET

A tuck placed in each side of a tube of blown tubing as produced to provide a convenient square or rectangular package, similar to that of the familiar brown paper bag or sack, in subsequent packaging.

GUTTA-PERCHA

A rubber-like material obtained from the leaves and bark of certain tropical trees. Sometimes used for the insulation of electrical wiring, and for transmission belting and various adhesives.

H

HALOCARBON PLASTICS

Plastics based on resins made by the polymerization of monomers composed only of carbon and a halogen or halogens.

HARDENER

A substance or mixture of substances added to plastic composition, or an adhesive to promote or control the curing reaction by taking part in it. The term is also used to designate a substance added to control the degree of hardness of the cured film. See also *Catalyst*.

HARDNESS

The resistance of a plastics material to compression and indentation. Among the most important methods of testing this property are *Brinell hardness*, *Rockwell hardness* and *Shore hardness*, q.v.

HAZE

The degree of cloudiness in a plastics material.

HEAD

The end section of a blow molding machine (in a general extruder) in which the melt is transformed into a hollow parison.

HEAD SPACE

The space between the fill level of a container and the sealing plane.

HEAT DEFLECTION TEMPERATURE

See *Heat-Distortion Point*.

HEAT-DISTORTION POINT

The temperature at which a standard test bar (ASTM D648) deflects 0.010 in. under a stated load of either 66 or 264 p.s.i.

HEAT-SEALING

A method of joining plastic films by simultaneous application of heat and pressure to areas in contact. Heat may be supplied conductively or dielectrically.

HEATING CHAMBER

In injection molding, that part of the machine in which the cold feed is reduced to a hot melt. Also *Heating Cylinder*.

HEEL

The part of a container between the bottom bearing surface and the side wall.

HEEL RADIUS

The degree of curvature at the extreme bottom end of a container extending upward from the bearing surface. Also called *Base Radius*.

H.F. PREHEATING

See *Dielectric Heating*.

HIGH-LOAD MELT INDEX

The rate of flow of a molten resin through a 0.0825 inch orifice when subjected to a force of 21,600 grams at 190°C. See *Melt Index*.

HIGH POLYMER

A macromolecular substance which, as indicated by the term "polymer" and by the name (e.g., polyvinyl chloride) and formula (e.g., CH_2CHCL) by which it is identified, consists of molecules which are (at least approximately) multiples of the low molecular unit.

HIGH-PRESSURE LAMINATES

Laminates molded and cured at pressures not lower than 1000 p.s.i. and more commonly in the range of 1200 to 2000 p.s.i.

HOB

A master model in hardened steel used to sink the shape of a mold into a soft steel block.

HOBBING

Forming multiple mold cavities by forcing a hob q.v., into soft steel (for beryllium-copper) cavity blanks.

HOMOPOLYMER

A polymer, consisting of (neglecting the ends, branch junctions, and other minor irregularities) a single type of repeating unit.

HONEYCOMB

Manufactured product consisting of sheet metal or a resin impregnated sheet material (paper, fibrous glass, etc.) which has been formed into hexagonal-shaped cells. Used as core material for sandwich constructions.

HOOP STRESS

The force per unit area in the wall of the pipe in the circumferential orientation due to internal hydrostatic pressure.

HOPPER

Conical feed reservoir into which molding powder is loaded and from which it falls into a molding machine or extruder, sometimes through a metering device.

HOPPER DRYER

A combination feeding and drying device for extrusion and injection molding of thermoplastics. Hot air flows upward through the hopper containing the feed pellets.

HOPPER LOADER

A curved pipe through which molding powders are pneumatically conveyed from shipping drums to machine hoppers.

HOT GAS WELDING

A technique of joining thermoplastic materials (usually sheet) whereby the materials are softened by a jet of hot air from a welding torch, and joined together at the softened points. Generally a thin rod of the same material is used to fill and consolidate the gap.

HOT-RUNNER MOLD

A mold in which the runners are insulated from the chilled cavities and are kept hot. Parting line is at gate of cavity, runners are in separate plate(s), so they are not, as is the case usually, ejected with the piece.

HOT-STAMPING

Engraving operation for marking plastics in which roll leaf is stamped with heated metal dies onto the face of the plastics. Ink compounds can also be used. By means of felt rolls, ink is applied to type and by means of heat and pressure, type is impressed into the material, leaving the marking compound in the indentation.

HYDRAULIC

A system in which energy is transferred from one place to another by means of compression and flow of a fluid (e.g., water, oil).

HYDROCARBON PLASTICS

Plastics based on resins made by the polymerization of monomers composed of carbon and hydrogen only.

HYDROGENATION

Chemical process whereby hydrogen is introduced into a compound.

HYDROLYSIS

Chemical decomposition of a substance involving the addition of water.

HYGROSCOPIC

Tending to absorb moisture.

I

IMMISCIBLE

Descriptive of two or more fluids which are not mutually soluble.

IMPACT BAR (Specimen)

A test specimen of specified dimensions which is utilized to determine the relative resistance of a plastic to fracture by shock.

IMPACT RESISTANCE

Relative susceptibility of plastics to fracture by shock, e.g., as indicated by the energy expended by a standard pendulum type impact machine in breaking a standard specimen in one blow.

IMPACT STRENGTH

(1) The ability of a material to withstand shock loading. (2) The work done in fracturing, under shock loading, a specified test specimen in a specified manner.

IMPREGNATION

The process of thoroughly soaking a material such as wood, paper or fabric, with a synthetic resin so that the resin gets within the body of the material. The process is usually carried out in an impregnator.

IMPULSE SEALING

A heat sealing technique in which a pulse of intense thermal energy is applied to the sealing area for a very short time, followed immediately by cooling. It is usually accomplished by using an RF heated metal bar which is cored for water cooling or is of such a mass that it will cool rapidly at ambient temperatures.

INFRA-RED

Part of the electromagnetic spectrum between the visible light range and the radar range. Radiant heat is in this range, and infra-red heaters are much used in sheet thermoforming.

INHIBITOR

A substance that slows down chemical reaction. Inhibitors are sometimes used in certain types of monomers and resins to prolong storage life.

INJECTION BLOW MOLDING

A blow molding process in which the parison to be blown is formed by injection molding.

INJECTION MOLD

A mold into which a plasticated material is introduced from an exterior heating cylinder.

INJECTION MOLDING

A molding procedure whereby a heat-softened plastic material is forced from a cylinder into a relatively cool cavity which gives the article the desired shape.

INJECTION MOLDING CYCLE

The complete time cycle of operation utilized in injection molding of an object including injection; die close and die open time.

INJECTION PRESSURE

The pressure on the face of the injection ram at which molding material is injected into a mold. It is usually expressed in p.s.i.

INJECTION RAM

The ram which applies pressure to the plunger in the process of injection molding or transfer molding.

INORGANIC PIGMENTS

Natural or synthetic metallic oxides, sulfides, and other salts, calcined during processing at 1200 to 2100°F. They are outstanding in heat- and light-stability, weather resistance, and migration resistance.

INSERT

An integral part of a plastics molding consisting of metal or other material which may be molded into position or may be pressed into the molding after the molding is completed.

IN SITU FOAMING

The technique of depositing a foamable plastics (prior to foaming) into the place where it is intended that foaming shall take place. An example is the placing of foamable plastics into cavity brickwork to provide insulation. After being positioned, the liquid mix foams to fill the cavity. See also *Foamed Plastics.*

INSTRON

An instrument utilized to determine the tensile and compressive properties of material.

INSULATION

A coating of a dielectric or essentially non-conducting material whose purpose it is to prevent the transmission of electricity.

INSULATION RESISTANCE

The electrical resistance of an insulating material to a direct voltage. It is determined by measuring the leakage of current which flows through the insulation.

INTERLOCK

A safety device designed to insure that a piece of apparatus will not operate until certain precautions have been taken.

INTERNAL MIXERS

Mixing machines using the principle of cylindrical containers in which the materials are deformed by rotating blades or rotors. The containers and rotors are cored so that they can be heated or cooled to control the temperature of a batch. These mixers are extensively used in the compounding of plastics and rubber materials and have the inherent advantage of keeping dust and fume hazards to a minimum.

INTRINSIC VISCOSITY

The intrinsic viscosity of a polymer is the limiting value of infinite dilution of the ratio at the specific viscosity of the polymer solution to its concentration in moles per liter.

$$n_1 = \underset{C \to 0}{\text{Lim}} \ \frac{n^{sp}}{C}$$

where n_1 = intrinsic viscosity, n^{sp} = specific viscosity, C = concentration in moles per liter. Intrinsic viscosity is usually estimated by determining the specific viscosity at several low concentrations and extrapolating the values of $\frac{n^{sp}}{C}$ to $C = O$. The concentration is expressed in terms of the repeating unit. In the case of polystyrene the repeating unit is $- CH_2 - CH(C_6H_5) -$ and has a molecular weight of 104.

INTROFACTION

The change in fluidity and wetting properties of an impregnating material, produced by the addition of an introfier, q.v.

INTROFIER

A chemical which will convert a colloidal solution into a molecular one. See also *Introfaction*.

ION EXCHANGE RESINS

Small granular or bead-like particles containing acidic or basic groups, which will trade ions with salts in solutions. Generally used for softening and purifying water.

IONOMER RESINS

A new polymer which has ethylene as its major component, but containing both covalent and ionic bonds. The polymer exhibits very strong interchain ionic forces. The anions hang from the hydrocarbon chain and the cations are metallic — sodium, potassium magnesium. These resins have many of the same features as polyethylene plus high transparency, tenacity, resilience and increased resistance to oils, greases and solvents. Fabrication is carried out as with polyethylene.

IRRADIATION (ATOMIC)

As applied to plastics, refers to bombardment with a variety of subatomic particles, generally alpha-, beta-, or gamma-rays.

Atomic irradiation has been used to initiate polymerization and co-polymerization of plastics and in some cases to bring about changes in the physical properties of a plastic material.

ISINGLASS
A white, tasteless gelatine derived from the bladder of fishes, usually the sturgeon. It is used as an adhesive and clarifying agent.

ISO
International Organization of Standardization.

ISOCYANATE RESINS
Most applications for this resin are based on its combination with polyols (e.g., polyesters, poly-ethers, etc.). During this reaction, the reactants are joined through the formation of the urethane linkage — and hence this field of technology is generally known as urethane chemistry.

ISOTACTIC
Pertaining to a type of polymeric molecular structure containing a sequence of regularly spaced asymmetric atoms arranged in like configuration in a polymer chain.

IZOD IMPACT TEST
A test designed to determine the resistance of a plastics material to a shock loading. It involves the notching of a specimen, which is then placed in the jaws of the machine and struck with a weighted pendulum. See also *Impact Strength*.

J

JACKET
A tough sheath to protect an insulated wire or cable, or to permanently group two or more insulated wires or cables.

JET MOLDING
Processing technique characterized by the fact that most of the heat is applied to the material as it passes through the nozzle or jet, rather than in a heating cylinder as is done in conventional processes.

JET SPINNING
For most purposes similar to melt spinning. Hot gas jet spinning uses a directed blast or jet of hot gas to "pull" molten polymer from a die lip and extend it into fine fibers.

JETTING
Turbulent flow of resin from an undersize gate or thin section into a thicker mold section, as opposed to lamular flow of material progressing radially from a gate to the extremities of the cavity.

JIG
Tool for holding component parts of an assembly during the manufacturing process, or for holding other tools. Also called a *Fixture*.

JUTE
Bast fiber obtained from the stems of several species of the plant Corchorus found mainly in India and Pakistan. Used as a filler, q.v., for plastics molding materials, and more recently as a reinforcement for polyester resins in the fabrication of reinforced plastics.

K

KIRKSITE
An alloy of aluminum and zinc used for the construction of blow molds; it imparts high degree of heat conductivity to the mold.

KISS-ROLL COATING
This roll arrangement carries a metered film of coating to the web; at the line of web contact, it is split with part remaining on the

roll, the remainder of the coating adhering to the web.

KNIFE COATING

A method of coating a substrate (usually paper or fabric) in which the substrate, in the form of a continous moving web, is coated with a material whose thickness is controlled by an adjustable knife or bar set at a suitable angle to the substrate. In the plastics industry PVC formulations are widely used in this work and curing is effected by passing the coated substrate into a special oven, usually heated by infrared lamps or convected air. There are a number of variations of this basic technique and they vary according to the type of product required.

KNIT LINES
See *Weld Mark.*

KNOCKOUT PIN
A device for knocking a cured piece from a mold. Also called *Ejector Pin.*

KNOT TENACITY (Knot Strength)
The tenacity in grams per denier of a yarn where an overhand knot is put into the filament or yarn being pulled to show up sensitivity to compressive or shearing forces.

KRAFT PAPER
Paper made from sulfate wood pulp.

L

LABEL PANEL
The plain portion of a decorated container set up for application of labels.

LACQUER
Solution of natural or synthetic resins, etc., in readily evaporating solvents, which is used as a protective coating.

LAMINAR FLOW
Laminar flow of thermoplastic resins in a mold is accompanied by solidification of the layer in contact with the mold surface that acts as an insulating tube through which material flows to fill the remainder of the cavity. This type of flow is essential to duplication of the mold surface.

LAMINATE
A product made by bonding together two or more layers of material or materials.

LAMINATED PLASTICS (Synthetic Resin-Bonded Laminate,
A plastics material consisting of superimposed layers of a synthetic resin-impregnated or -coated filler which have been bonded together, usually by means of heat and pressure, to form a single piece.

LAMINATED WOOD
A high-pressure bonded wood product composed of layers of wood with resin as the laminating agent. The term plywood covers a form of laminated wood in which successive layers of veneer are ordinarily cross laminated, the core of which may be veneer or sawn lumber in one piece or several pieces.

LAND
(1) The horizontal bearing surface of a semipositive or flash mold by which excess material escapes. See *Cut-off.* (2) The bearing surface along the top of the flights of a screw in a screw extruder. (3) The surface of an extrusion die parallel to the direction of melt flow.

LATTICE PATTERN

In reinforced plastics, a pattern of filament winding with a fixed arrangement of open voids.

LAY-UP

(n.) As used in reinforced plastics, the reinforcing material placed in position in the mold; also the resin-impregnated reinforcement. (v.) — The process of placing the reinforcing material in position in the mold.

L/D RATIO

A term used to define an extrusion screw which denotes the ratio of the screw length to the screw diameter.

LEACH

To extract a soluble component from a mixture by the process of percolation.

LEAKER

Any condition of the finish where the normal sealing device or closure will not retain the air or liquid content of the bottle.

LIGHT-RESISTANCE

The ability of a plastics material to resist fading after exposure to sunlight or ultra-violet light. Nearly all plastics tend to darken under these conditions.

LIGNIN PLASTICS

Plastics based on lignin resins.

LIGNIN RESIN

A resin made by heating lignin or by reaction of lignin with chemicals or resins, the lignin being in greatest amount of mass.

LIMITING OXYGEN INDEX (LOI)

The concentration of oxygen required to maintain burning. See ASTM Procedure D2863-74.

LINEAR MOLECULE

A long chain molecule as contrasted to one having many side chains or branches.

LINTERS

Short fibers that adhere to the cotton seed after ginning. Used in rayon manufacture, as fillers for plastics, and as a base for the manufacture of cellulosic plastics.

LIP

The extreme outer edge of the top of a container intended to facilitate pouring.

LITHARGE

PbO. An oxide of lead used as an inorganic accelerator, as a vulcanizing agent for neoprene, and as an ingredient of paints.

LOADING TRAY (Charging Tray)

A device in the form of a specially designed tray which is used to load the charge simultaneously into each cavity of a multi-cavity mold by the withdrawal of a sliding bottom from the tray.

LOOP TENACITY (Loop Strength)

The tenacity or strength value obtained by pulling two loops, as two links in a chain, against each other to demonstrate the susceptibility that a yarn, cord or rope has for cutting or crushing itself.

LOSS FACTOR

The product of the power factor and the dielectric constant.

LOW PRESSURE LAMINATES

In general, laminates molded and cured in the range of pressures from 400 p.s.i. down to and including pressures obtained by the mere contact of the plies.

LUBRICANTS, SOLID

A solid substance that will reduce the friction or prevent sticking when placed between two moving parts, i.e., graphite, molybdenum disulfide, etc.

LUG

An indentation or raised portion of the surface of a container, provided to control automatic (multicolor) decorating operations.

LUMINESCENT PIGMENTS

Special pigments available to produce striking effects in the dark. Basically there are two types: one is activated by ultra-violet radiation, producing very strong luminescence and, consequently, very eyecatching effects; the other type, known as phosphorescent pigments, does not require any separate source of radiation.

LUMINOUS TRANSMITTANCE

The ratio of the luminous flux transmitted by a body to the flux incident upon it.

M

MACERATE

(v.) To chop or shread fabric for use as a filler for a molding resin. (n.) The molding compound obtained when so filled.

MACHINE SHOT CAPACITY

Refers to the maximum weight of thermoplastic resin which can be displaced or injected by the injection ram in a single stroke.

MACROMOLECULE

The large ("giant") molecules which make up the high polymers.

MANDREL

(1) The core around which paper, fabric, or resin-impregnated fibrous glass is wound to form pipes or tubes. (2) In extrusion, the central finger of a pipe or tubing die.

MANIFOLD

A term used mainly with reference to blow molding and sometimes with injection molding equipment. It refers to the distribution or piping system which takes the single channel flow output of the extruder or injection cylinder and divides it to feed several blow molding heads or injection nozzles.

MASTERBATCH

A plastics compound which includes a high concentration of an additive or additives. Masterbatches are designed for use in appropriate quantities with the basic resin or mix so that the correct end concentration is achieved. For example, color masterbatches for a variety of plastics are extensively used as they provide a clean and convenient method of obtaining accurate color shades.

MAT

A randomly distributed felt of glass fibers used in reinforced plastics lay-up molding.

MATERIAL WELL

Space provided in a compression or transfer mold to care for bulk factor.

MATCHED METAL MOLDING

Method of molding reinforced plastics between two close-fitting metal molds mounted in a hydraulic press.

MELAMINE PLASTICS

Thermosetting plastics made from melamine and formaldehyde resins.

MELT EXTRACTOR

Usually refers to a type of injection machine torpedo but could refer to any type of device which is placed in a plasticating system for the purpose of separating fully plasticated melt from partially molten pellets and material. It thus insures a fully plasticated discharge of melt from the plasticating system.

MELT FLOW

The flow rate obtained from extrusion of a molten resin through a die of specified length and diameter under prescribed conditions of time, temperature and load as set forth in ASTM D1238.

MELT FRACTURE

An instability in the melt flow through a die starting at the entry to the die. It leads to surface irregularities on the finished article like a regular helix or irregularly-spaced ripples.

MELT INDEX

The amount, in grams, of a thermoplastic resin which can be forced through a 0.0825 inch orifice when subjected to 2160 gms. force in 10 minutes at $190°C$.

MELTING POINT

The temperature at which solid and liquid forms of a substance are in equilibrium. In common usage the melting point is taken as the temperature at which the liquid first forms in a small sample as its temperature is increased gradually.

MELT INSTABILITY

An instability in the melt flow through a die starting at the land of the die. It leads to the same surface irregularities on the finished part as melt fracture.

MELT STRENGTH

The strength of the plastic while in the molten state.

MELT TEMPERATURE

The temperature of the molten plastic just prior to entering the mold or extruded through the die.

MENISCUS

The free surface of a liquid in a container, for example, water in contact with air confined in a capillary tube. The meniscus may be convex, e.g., mercury vs. air in glass, or concave, e.g., water vs. air in glass.

MER

The repeating structural unit of any high polymer.

METALIZING

Applying a thin coating of metal to a nonmetallic surface. May be done by chemical deposition or by exposing the surface to vaporized metal in a vacuum chamber.

METALLIC PIGMENTS

A class of pigments consisting of thin opaque aluminum flakes (made by ball milling either a disintegrated aluminum foil or a rough metal powder and then polishing to obtain a flat, brilliant surface on each particle) or copper alloy flakes (known as bronze pigments). Incorporated into plastics, they produce unusual silvery and other metal-like effects.

METALLIC "RING"

The audible sound emitted when most metals are dropped upon a hard surface.

METERING SCREW

An extrusion screw which has a shallow constant depth, and constant pitch section over, usually, the last 3 to 4 flights.

METHYL METHACRYLATE

$CH_2CCH_3COOCH_3$. A colorless, volatile liquid derived from acetone cyanohydrin, methanol and dilute sulphuric acid, and used in the production of acrylic resins, q.v.

MIGRATION OF PLASTICIZER

Loss of plasticizer from an elastomeric plastic compound with subsequent absorption by an adjacent medium of lower plasticizer concentration.

Mn (Number-average molecular weight)

The total weight of all molecules divided by the total number of molecules.

MODIFIED

Containing ingredients such as fillers, pigments or other additives, that help to vary the physical properties of a plastics material. An example is oil modified resin.

MODULUS OF ELASTICITY

The ratio of stress to strain in a material that is elastically deformed.

MOISTURE VAPOR TRANSMISSION

The rate at which water vapor permeates through a plastic film or wall at a specified temperature and relative humidity.

MOLD

(v.) To shape plastic parts or finished articles by heat and pressure. (n.) (1) The cavity or matrix into which the plastic composition is placed and from which it takes its form. (2) The assembly of all the parts that function collectively in the molding process.

MOLD EFFICIENCY

In a multimold blowing system the percentage of the total turn-around time of the mold actually required for forming, cooling and ejection of the container.

MOLDING CYCLE

(1) The period of time occupied by the complete sequence of operations on a molding press requisite for the production of one set of moldings. (2) The operations necessary to produce a set of moldings without reference to the time taken.

MOLDING POWDER

Plastic material in varying stages of granulation, and comprising resin, filler, pigments, plasticizers, and other ingredients, ready for use in the molding operation.

MOLDING PRESSURE

The pressure applied to the ram of an injection machine or press to force the softened plastic completely to fill the mold cavities.

MOLDING SHRINKAGE (Mold Shrinkage, Shrinkage, Contraction)

The difference in dimensions, expressed in inches per inch, between a molding and the mold cavity in which it was molded, both the mold and the molding being at normal room temperature when measured.

MOLD RELEASE

See *Parting Agent.*

MOLD SEAM

A vertical line formed at the point of contact of the mold halves. The prominence of the line depends on the accuracy with which the mating mold halves are matched. See *Parting Line.*

MOLECULAR WEIGHT DISTRIBUTION

The ratio of the weight average molecular weight to the number average molecular weight gives an indication of the distribution.

"MOLY"

Common abbreviation used to denote a compound containing molybdenum.

MONOFILAMENT (Monofil)

A single filament of indefinite length. Monofilaments are generally produced by extrusion. Their outstanding uses are in the fabrication of bristles, surgical sutures, fishing leaders, tennis-racquet strings, screen materials, ropes and nets; the finer monofilaments are woven and knitted on textile machinery.

MONOMER

A relatively simple compound which can react to form a polymer. See also *Polymer*.

MOUNTING PLATE

The part of the blow molding unit to which the mold is attached.

MOVABLE PLATEN

The large back platen of an injection molding machine to which the back half of the mold is secured during operation. This platen is moved either by a hydraulic ram or a toggle mechanism.

MULTI-CAVITY MOLD

A mold with two or more mold impressions, i.e., a mold which produces more than one molding per molding cycle.

MULTIFILAMENT YARN

The multifilament yarn is composed of a multitude of fine continuous filaments, often 5 to 100 individual filaments, usually with some twist in the yarn to facilitate handling. Multifilament yarn sizes are described in denier and range from 5-10 denier up to a few hundred denier. The larger deniers, even in the thousands, are usually obtained by plying smaller yarns

together. Individual filaments in a multifilament yarn are usually about 1 to 5 denier (which is about ½ mil to 1 mil diameter in Marlex polyethylene).

MULTIPLE HEAD MACHINE

A (blow molding) machine in which the plastic melt prepared by the extruder is divided into a multiplicity of separate streams (parisons) each giving ultimately a finished item.

Mw (Weight-average molecular weight)

The sum of the total weights of molecules of each size multiplied by their respective weights divided by the total weight of all molecules.

N

NECK

The part of a container where the shoulder cross section area decreases to form the finish.

NECK BEAD

A protruding circle on a container at the point where the neck meets the finish, the diameter of which usually equals the outside diameter of the closure.

NECK-IN

In extrusion coating, the difference between the width of the extruded web as it leaves the die and the width of the coating on the substrate.

NECK INSERT

Part of the mold assembly which forms the neck and finish. Sometimes called Neck Ring.

NEEDLE BLOW

A specific blow molding technique where the blowing air is injected into the hollow article through a sharpened hollow needle which pierces the parison.

NEST PLATE
A retainer plate with a depressed area for cavity blocks used in injection molding.

NIP
The "V" formed where the pressure roll contacts the chill roll.

NONPOLAR
Having no concentrations of electrical charge on a molecular scale, thus, incapable of significant dielectric loss. Examples among resins are polystyrene and polyethylene.

NONRIGID PLASTIC
A non-rigid plastic is one which has a stiffness or apparent modulus of elasticity of not over 50,000 p.s.i. at 25°C when determined according to ASTM test procedure D747-43 T.

NOTCH SENSITIVITY
The extent to which the sensitivity of a material to fracture is increased by the presence of a surface in homogenity such as a notch, a sudden change in section, a crack, or a scratch. Low notch sensitivity is usually associated with ductile materials, and high notch sensitivity with brittle materials.

NOVOLAC
A phenolic-aldehyde resin which, unless a source of methylene groups is added, remains permanently thermoplastic. See also *Resinoid* and *Thermoplastic*.

NOZZLE
The hollow cored metal nose screwed into the extrusion end of (a) the heating cylinder of an in-

jection machine or (b) a transfer chamber where this is a separate structure. A nozzle is designed to form under pressure a seal between the heating cylinder or the transfer chamber and the mold. The front end of a nozzle may be either flat or spherical in shape.

NYLON
The generic name for all synthetic fiber-forming polyamides; they can be formed into monofilaments and yarns characterized by great toughness, strength and elasticity, high melting point; and good resistance to water and chemicals. The material is widely used for bristles in industrial and domestic brushes, and for many textile applications; it is also used in injection molding gears, bearings, combs, etc.

O

OBLONG
A particular shape. A container which has a rectangular cross section perpendicular to the major axis.

OFF CENTER
Any condition where the finish opening is not centered over the bottom of the container. Also, the condition where the mandrel is not concentric with the ring of the blowing head.

OFFSET
A printing technique in which ink is transferred from a bath onto the raised surface of the printing plate by rollers. Subsequently, the printing plates transfer the ink to the object to be printed.

OIL-SOLUBLE RESIN
Resin which at moderate temperatures will dissolve in, disperse in,

or react with, drying oils to give a homogeneous film of modified characteristics.

OLEFINS

A group of unsaturated hydro-carbons of the general formula CnH_2n, and named after the corresponding paraffins by the addition of "ene" or "ylene" to the stem. Examples are ethylene and propylene.

OLEO RESINS

Semi-solid mixtures of the resin and essential oil of the plant from which they exude, and sometimes referred to as balsams. Oleo-resinous materials also consist of products of drying oils and natural or synthetic resins.

OLIGOMER

A polymer consisting of only a few monomer units such as a dimer, trimer, tetramer, etc., or their mixtures.

ONE-SHOT MOLDING

In the urethane foam field, indi-cates a system whereby the iso-cyanate, polyol, catalyst, and other additives are mixed together directly and a foam is produced immediately (as distinguished from *Prepolymer*).

OPAQUE

Descriptive of a material or sub-stance which will not transmit light. Opposite of transparent, q.v. Materials which are neither opaque nor transparent are some-times described as semi-opaque, but are more properly classified as translucent, q.v.

OPEN-CELL FOAMED PLASTIC

A cellular plastic in which there is a predominance of interconnected cells.

ORANGE-PEEL

Said of injection moldings that have unintentionally rough sur-faces.

ORGANIC PIGMENTS

Characterized by good brightness and brilliance. They are divided into toners and lakes. Toners, in turn, are divided into insoluble organic toners and lake toners. The insoluble organic toners are usually free from salt-forming groups. Lake toners, are prac-tically pure, water-insoluble heavy metal salts of dyes without the fillers or substrates of ordinary lakes. Lakes, which are not as strong as lake toners, are water-insoluble heavy metal salts or other dye complexes precipitated upon or admixed with a base or filler.

ORGANOSOL

A vinyl or nylon dispersion, the liquid phase of which contains one or more organic solvents. See also *Plastisol.*

ORIENTATION

The alignment of the crystalline structure in polymeric materials so as to produce a highly uniform structure. Can be accomplished by cold drawing or stretching during fabrication.

ORIFICE

The opening in the extruder die formed by the orifice bushing (ring) and mandrel.

ORIFICE BUSHING

The outer part of the die in an extruder head.

OUT-OF-ROUND

A plastic container manufacturing variance in which a round con-tainer, when formed, does not remain round.

OVAL

A particular shape. A container which has an egg-shaped cross section perpendicular to the major axis.

OVERCOATING

In extrusion coating, the practice of extruding a web beyond the edge of the substrate web.

OVERFLOW CAPACITY

The capacity of a container to the top of the finish or to the point of overflow.

OVERLAY SHEET (Surfacing Mat)

A nonwoven fibrous mat (either in glass, synthetic fiber, etc.) used as the top layer in a cloth or mat layup to provide a smoother finish or minimize the appearance of the fibrous pattern.

OXIDATION

The addition of oxygen to a compound or the reduction of hydrogen.

OXYGEN INDEX

See *Limiting Oxygen Index.*

P

PANELING

Distortion of a container occurring during aging or storage, caused by the development of a reduced pressure inside the container.

PARALLELS

(1) Spacers placed between the steam plate and press platen to prevent the middle section of the mold from bending under pressure. (2) Pressure pads or spacers between the steam plates of a mold to control height when closed and to prevent crushing the parts of the mold when the land area is inadequate.

PARISON

The hollow plastic tube from which a container, toy, etc. is blow molded.

PARISON SWELL

In blow molding the ratio of the cross-sectional area of the parison to the cross-sectional area of the die opening.

PARTING AGENT

A lubricant, often wax, used to coat a mold cavity to prevent the molded piece from sticking to it, and thus to facilitate its removal from the mold. Also called *Release Agent.*

PARTING LINE

Mark on a molding or casting where halves of mold met in closing.

PARTITIONED MOLD COOLING

A large diameter hole drilled into the mold (usually the core) and partitioned by a metal plate extending to near the bottom end of the channel. Water is introduced near the top of one side of the partition and removed on the other side.

PEARLESCENT PIGMENTS

A class of pigments consisting of particles that are essentially transparent crystals of a high refractive index. The optical effect is one of partial reflection from the two sides of each flake. When reflections from parallel plates reinforce each other, the result is a silvery luster. Effects possible range from brilliant highlighting to moderate enhancement of the normal surface gloss.

PELLET

A small ball or spherical shape.

PELLETIZING

A process of producing pellets.

PERMANENT SET

The increase in length, expressed in a percentage of the original length, by which an elastic material fails to return to original length after being stressed for a standard period of time.

PERMEABILITY

(1) The passage or diffusion of a gas, vapor, liquid, or solid through a barrier without physically or chemically affecting it. (2) The rate of such passage.

pH

An expression of the degree of acidity or alkalinity of a substance. Neutrality is pH_7 — acid solutions being under 7 and alkaline solutions over 7. pH meters are commercially available for accurate readings.

PHENOLIC RESIN

A synthetic resin produced by the condensation of an aromatic alcohol with an aldehyde, particularly of phenol with formaldehyde. Phenolic resins form the basis of thermosetting molding materials, laminated sheet, and stoving varnishes. They are also used as impregnating agents and as components of paints, varnishes, lacquers, and adhesives.

PHENOXY RESINS

A high molecular weight thermoplastic polyester resin based on bisphenol-A and epichorohydrin. Recently developed in the United States, the material is available in grades suitable for injection molding, extrusion, coatings and adhesives, q.v.

PHTHALATE ESTERS

A main group of plasticizers, q.v., produced by the direct action of alcohol on phthalic anhydride. The phthalates are the most widely used of all plasticizers, and are generally characterized by moderate cost, good stability, and good all-round properties.

PHTHALOCYANINE PIGMENTS

Organic pigments, q.v., of extremely stable chemical configuration resulting in very good fastness properties. These properties are enhanced by the formation of the copper complex which is the phthalocyanine blue most used. The introduction of chlorine atoms into the molecule of blue gives the well-known phthalocyanine green, also usually in the form of copper complex.

PIGMENT

Any colorant, usually an insoluble powdered substance used to produce a desired color or hue.

PILL

See *Preform*.

PIMPLE

An imperfection, a small, sharp, or conical elevation on the surface of a plastic product.

PINCH-OFF

A raised edge around the cavity in the mold which seals off the part and separates the excess material as the mold closes around the parison in the blow molding operation.

PINCH-OFF BLADES

The part of the mold which compresses the parison to effect sealing of the parison prior to blowing and to permit easy removal and cooling of flash.

PINCH-OFF LAND

The width of pinch-off blade which effects sealing of the parison.

PINCH-OFF TAIL

The bottom of the parison that is pinched off when the mold closes.

PINHOLE

A very small hole in the extruded resin coating.

PINPOINT GATE

A restricted orifice of 0.030 inches or less in diameter through which molten resin flows into a mold cavity.

PIPE TRAIN

A term used in extrusion of pipe which denotes the entire equipment assembly used to fabricate the pipe, e.g., extruder, die, cooling bath, haul-off and cutter.

PIT

An imperfection, a small crater in the surface of the plastic, with its width of approximately the same order of magnitude as its depth.

PITCH

The distance from any point on the flight of a screw line to the corresponding point on an adjacent flight, measured parallel to the axis of the screw line or threading.

PLASTIC

(n.) One of many high-polymeric substances, including both natural and synthetic products, but excluding the rubbers. At some stage in its manufacture every plastic is capable of flowing, under heat and pressure, if necessary, into the desired final shape. (v.) Made of plastic; capable of flow under pressure or tensile stress.

PLASTIC DEFORMATION

A change in dimensions of an object under load that is not recovered when the load is removed; opposed to elastic deformation.

PLASTIC MEMORY

A phenomenon of plastics to return to its original molded form. Different plastics possess varying degrees of this characteristic.

PLASTICS TOOLING

Tools, e.g., dies, jigs, fixtures, etc., for the metal forming trades constructed of plastics, generally laminates or casting materials.

PLASTICATE

To soften by heating or kneading. Synonyms are: plastify, flux, and, (imprecisely) plasticize q.v.

PLASTICITY

The quality of being able to be shaped by plastic flow.

PLASTICIZE

To soften a material and make it plastic or moldable, either by means of a plasticizer or the application of heat.

PLASTICIZER

Chemical agent added to plastic compositions to make them softer and more flexible.

PLASTIGEL

A plastisol exhibiting gel-like flow properties.

PLASTISOLS

Mixtures of resins and plasticizers which can be molded, cast, or converted to continuous films by the application of heat. If the mixtures contain volatile thinners also, they are known as *Organosols.*

PLASTOMETER

An instrument for determining the flow properties of a thermoplastic resin by forcing the molten resin through a die or orifice of specific size at a specified temperature and pressure.

PLATE DISPERSION PLUG

Two perforated plates held together with a connecting rod which are placed in the nozzle of an injection molding machine to aid in dispersing a colorant in a resin as it flows through the orifices in the plates.

PLATE-MARK
Any imperfection in a pressed plastic sheet resulting from the surface of the pressing plate.

PLATENS
The mounting plates of a press to which the entire mold assembly is bolted.

PLATFORM BLOWING
A special technique for blowing large parts. To prevent excessive sag of the heavy parison the machine employs a table which after rising to meet the parison at the die descends with the parison but at a slightly lower rate than the parison extrusion speed.

PLUG-AND-RING
Method of sheet forming in which a plug, functioning as a male mold, is forced into a heated plastic sheet held in place by a clamping ring.

PLUG FORMING
A thermoforming process in which a plug or male mold is used to partially preform the part before forming is completed using vacuum or pressure.

PLUNGER
See *Force Plug.*

PNEUMATIC
A system in which energy is transferred by compression, flow and expansion of air.

POCK MARKS
Irregular indentations on the surface of a blown container caused by insufficient contact of the blown parison with the mold surface. They are due to low blow pressure, air gas entrapment or moisture condensation on mold surface.

POISE
The unit of viscosity, q.v., expressed as one dyne per second per square centimeter.

POLAR
See *Nonpolar.*

POLISHING ROLL(s)
A roll or series of rolls, which have a highly polished chrome plated surface, that are utilized to produce a smooth surface on sheet as it is extruded.

POLYACRYLATE
A thermoplastic resin made by the polymerization of an acrylic compound such as methyl methacrylate.

POLYALLOMERS
Crystalline polymers produced from two or more olefin monomers.

POLYAMIDE
A polymer in which the structural units are linked by amide or thioamide groupings. Many polyamides are fiber-forming.

POLYBLENDS
Colloquial term generally used in the styrene field to apply to mechanical mixtures of polystyrene and rubber.

POLYBUTYLENE
A polymer prepared by the polymerization of butene as the sole monomer. (See *Polybutylene Plastics* and *Butylene Plastics.*)

POLYBUTYLENE PLASTICS
Plastics based on polymers made with butene as essentially the sole monomer.

POLYCARBONATE RESINS
Polymers derived from the direct reaction between aromatic and aliphatic dihydroxy compounds with phosgene or by the ester exchange reaction with appropriate phosgene-derived precursors.

POLYCONDENSATION
See *Condensation.*

POLYESTER
A resin formed by the reaction between a dibasic acid and a dihydroxy alcohol, both organic. Modification with multi-functional acids and/or bases and some unsaturated reactants permit crosslinking to thermosetting resins. Polyesters modified with fatty acids are called *Alkyds.*

POLYETHYLENE
A thermoplastic material composed by polymers of ethylene. It is normally a translucent, tough, waxy solid which is unaffected by water and by a large range of chemicals.

POLYIMIDE RESINS
A new group of resins recently introduced in the United States. The material is an aromatic polyimide made by reacting pyromellitic dianhydride with aromatic diamines. The polymer is characterized by the fact that it has rings of four carbon atoms tightly bound together, and the manufacturers claim that the new resin has greater resistance to heat than any other unfilled organic material yet discovered. Suggested applications include components for internal combustion engines.

POLYISOBUTYLENE
The polymerization product of isobutylene. It varies in consistency from a viscous liquid to a rubber-like solid with corresponding variation in molecular weight from 1000 to 400,000.

POLYLINER
A perforated longitudinally ribbed sleeve that fits inside the cylinder of an injection molding machine; used as a replacement for conventional injection cylinder torpedos.

POLYMER
A high-molecular-weight organic compound, natural or synthetic, whose structure can be represented by a repeated small unit, the *mer*; e.g., polyethylene, rubber, cellulose. Synthetic polymers are formed by addition or condensation polymerization of monomers. If two or more monomers are involved, a copolymer is obtained. Some polymers are elastomers, some plastics.

POLYMERIZATION
A chemical reaction in which the molecules of a monomer are linked together to form large molecules whose molecular weight is a multiple of that of the original substance. When two or more monomers are involved, the process is called copolymerization or heteropolymerization. See also *Degree of, Condensation,* and *Polymer.*

POLYMETHYL METHACRYLATE
A thermoplastic material composed of polymers of methyl methacrylate. It is a transparent solid with exceptional optical properties and good resistance to water. It is obtainable in the form of sheets, granules, solutions, and emulsions. It is extensively used for aircraft domes, lighting fixtures, decorative articles, etc.; it is also used in optical instruments and surgical appliances.

POLYOLEFIN

A polymer prepared by the polymerization of an *Olefin*(s) as the sole *Monomer*(s). (See *Polyolefin plastics*).

POLYOLEFIN PLASTICS

Plastics based on polymer made with an *Olefin*(s) as essentially the sole *Monomer*(s).

POLYOXYMETHYLENE

A polymer in which the repeated structural unit in the chain is oxymethylene.

POLYOXYMETHYLENE PLASTICS

Plastics based on polymers in which oxymethylene is essentially the sole repeated structural unit in the chains.

POLYPHENYLENE OXIDE

Presently made commercially as a polyether of 2, 6-dimethyl-phenol via an oxidative coupling process by means of air or pure oxygen in the presence of a copper-amine complex catalyst. These resins have a useful temperature range from less than -275°F to 375°F with intermittent use up to 400°F possible.

POLYPHENYLENE SULFIDE

A crystalline aromatic thermoplastic polymer with a symmetrical, rigid backbone chain consisting of para-substituted benzene rings connected by a single sulfur atom between rings. Exhibits extremely high heat and chemical resistance.

POLYPROPYLENE

A tough, lightweight rigid plastic made by the polymerization of high-purity propylene gas in the presence of an organometallic catalyst at relatively low pressures and temperatures.

POLYSTYRENE

A water-white thermoplastic produced by the polymerization of styrene (vinyl benzene). The electrical insulating properties of polystyrene are outstandingly good and the material is relatively unaffected by moisture. In particular the power loss factor is extremely low over the frequency range $10^3 - 10^8$ c.p.s.

POLYTERPENE RESINS

Thermoplastic resins obtained by the polymerization of turpentine in the presence of catalysts. These resins are used in the manufacture of adhesives, coatings, and varnishes, and in food packaging. They are compatible with waxes, natural and synthetic rubbers, and polyethylene.

POLYTETRAFLUORO-ETHYLENE (PTFE) RESINS

Members of the fluorocarbons, q.v., family of plastics made by the polymerization of tetrafluoroethylene. PTFE is characterized by its extreme inertness to chemicals, very high thermal stability and low frictional properties. Among the applications for these materials are bearings, fuel hoses, gaskets and tapes, and coatings for metal and fabric.

POLYURETHANE RESINS

A family of resins produced by reacting diisocyanate with organic compounds containing two or more active hydrogens to form polymers having free isocyanate groups. These groups, under the influence of heat or certain catalysts, will react with each other, or with water, glycols, etc., to form a thermosetting material.

POLYVINYL ACETAL

A member of the family of vinyl plastics, q.v., polyvinyl acetal is the general name for resins produced from a condensation of polyvinyl alcohol with an aldehyde. There are three main groups: polyvinyl acetal itself; polyvinyl butyral, and polyvinyl formal, q.v. Polyvinyl acetal resins are thermoplastics which can be processed by casting, extruding, molding and coating, but their main uses are in adhesives, lacquers, coatings and films.

POLYVINYL ACETATE

A thermplastic material composed of polymers of vinyl acetate in the form of a colorless solid. It is obtainable in the form of granules, solutions, latices, and pastes, and is used extensively in adhesives, for paper and fabric coatings, and in bases for inks and lacquers.

POLYVINYL ALCOHOL

A thermoplastic material composed of polymers of the hypothetical vinyl alcohol. Usually a colorless solid, insoluble in most organic solvents and oils, but soluble in water when the content of hydroxy groups in the polymer is sufficiently high.

The product is normally granular. It is obtained by the partial hydrolysis or by the complete hydrolysis of polyvinyl esters, usually by the complete hydrolysis of polyvinyl acetate. It is mainly used for adhesives and coatings.

POLYVINYL BUTYRAL

A thermoplastic material derived from a polyvinyl ester in which some or all of the acid groups have been replaced by hydroxyl groups and some or all of these hydroxyl groups replaced by butyral groups by reaction with butyraldehyde. It is a colorless flexible tough solid.

It is used primarily in interlayers for laminated safety glass.

POLYVINYL CARBAZOLE

A thermoplastic resin, brown in color, obtained by reacting acetylene with carbazole. The resin has excellent electrical properties and good heat and chemical resistance. It is used as an impregnant for paper capacitors.

POLYVINYL CHLORIDE (PVC)

A thermoplastic material composed of polymers of vinyl chloride; a colorless solid with outstanding resistance to water, alcohols, and concentrated acids and alkalies. It is obtainable in the form of granules, solutions, latices, and pastes. Compounded with plasticizers it yields a flexible material superior to rubber in aging properties. It is widely used for cable and wire coverings, in chemical plants, and in the manufacture of protective garments.

POLYVINYL CHLORIDE ACETATE

A thermoplastic material composed of copolymers of vinyl chloride and vinyl acetate; a colorless solid with good resistance to water, and concentrated acids and alkalies.

It is obtainable in the form of granules, solutions, and emulsions. Compounded with plasticizers it yields a flexible material superior to rubber in aging properties. It is widely used for cable and wire coverings, in chemicals plants, and in protective garments.

POLYVINYL FORMAL

One of the groups of polyvinyl acetal resins, q.v., made by the condensation of formaldehyde in the presence of polyvinyl alcohol. It is used mainly in combination with cresylic phenolics, for wire coatings and for impregnations, but can also be molded, extruded or cast. It is resistant to greases and oils.

POLYVINYLIDENE CHLORIDE
A thermoplastic material composed of polymers of vinylidene chloride (1,1-dichloroethylene). It is a white powder with softening temperature at 185-200°C. The material is also supplied as a copolymer with acrylonitrile or vinyl chloride, giving products which range from the soft flexible type to the rigid type. Also known as saran.

POLYVINYLIDENE FLUORIDE RESINS
This recent member of the fluorocarbons, q.v. family of plastics is a homopolymer of vinylidene fluoride. It is supplied as powders and pellets for molding and extrusion and in solution form for casting. The resin has good tensile and compressive strength and high impact strength. Among anticipated applications are chemical equipment such as gaskets, impellers and other pump parts, and packaging uses such as drum linings and protective coatings.

POROUS MOLDS
Molds which are made up of bonded or fused aggregate (powdered metal, coarse pellets, etc.) in such a manner that the resulting mass contains numerous open interstices of regular or irregular size through which either air or liquids may pass through the mass of the mold.

POSITIVE MOLD
A mold designed to trap all the molding material when it closes.

POSTCURE
Those additional operations to which a cured thermosetting plastic or rubber composition are subjected to enhance the level of one or more properties.

POSTFORMING
The forming, bending, or shaping of fully cured, C-stage thermoset laminates that have been heated to make them flexible. On cooling, the formed laminate retains the contours and shape of the mold over which it has been formed.

POT LIFE
See *Working Life*.

POTTING
Similar to *Encapsulating* q.v., except that steps are taken to insure complete penetration of all the voids in the object before the resin polymerizes.

POUR OUT FINISH
A container finish with an undercut below the top, designed to facilitate pouring without dripping.

POWDER MOLDING
General term used to denote several techniques for producing objects of varying sizes and shapes by melting polyethylene powder, usually against the inside of a mold. The techniques vary as to whether the molds are stationary (e.g., as in variations on slush molding techniques) or rotating (e.g., as in variations on rotational molding).

POWER FACTOR
In a perfect condenser, the current leads the voltage by 90°. When a loss takes place in the insulation, the absorbed current, which produces heat, throws the 90° relationship out according to the proportion of current absorbed by the dielectric. The power factor is the cosine of the angle between voltage applied and the current re-

sulting. Measurements are usually
made at million-cycle frequencies.

PREFORM

(*n.*) A compressed tablet or biscuit
of plastic composition used for
efficiency in handling and accuracy in weighing materials. (*v.*) To
make plastic molding powder into
pellets or tablets.

PREHEATING

The heating of a compound prior
to molding or casting in order to
facilitate the operation or to reduce the molding cycle.

PREHEAT ROLL

In extrusion coating, a heated roll
installed between the pressure roll
and unwind roll whose purpose is
to heat the substrate before it is
coated.

PREIMPREGNATION

The practice of mixing resin and
reinforcement before shipping it
to the molder.

PREMIX

In reinforced plastics molding, the
material made by "do-it-
yourselfers," molders, or end-users
who purchase polyester or phenolic resin, reinforcement, fillers,
etc., separately and mix the reinforced molding compounds on
their own premises.

PREPLASTICATION

Technique of premelting injection
molding powders in a separate
chamber, then transferring the
melt to the injection cylinder. Device used for preplastication is
commonly known as a preplas-
ticizer. See *Plasticate.*

PREPLASTICIZE (Preplasticizer)

See *Preplastication.*

PREPOLYMER MOLDING

In the urethane foam field, indicates a system whereby a portion
of the polyol is pre-reacted with
the isocyanate to form a liquid
propolymer with a viscosity range
suitable for pumping or metering.
This component is supplied to
end-users with a second premixed
blend of additional polyol, catalyst, blowing agent, etc. When the
two components are mixed together, foaming occurs. (See *One-
Shot Molding*).

PREPREG

A term generally used in reinforced plastics to mean the reinforcing material containing or
combined with the full complement of resin before molding.

PREPRINTING

In sheet thermoforming, the distorted printing of sheets before
they are formed. During forming
the print assumes its proper proportions.

PRESS POLISH

A finish for sheet stock produced
by contact, under heat and pressure, with a very smooth metal
which gives the plastic a high
sheen.

PRESSURE BREAK

As applied to a defect in a laminated plastic a break apparent in
one or more outer sheets of the
paper, fabric, or other base visible
through the surface layer of resin
which covers it.

PRESSURE FORMING

A thermoforming process wherein
pressure is used to push the sheet
to be formed against the mold surface as opposed to using a vacuum
to suck the sheet flat against the
mold.

PRESSURE PADS

Reinforcements of hardened steel distributed around the dead areas in the faces of a mold to help the land absorb the final pressure of closing without collapsing.

PRESSURE ROLL

In extrusion coating, the roll which with the chill roll applies pressure to the substrate and the molten extruded web.

PRESSURE SENSITIVE ADHESIVE

An adhesive which develops maximum bonding power by applying only a light pressure.

PRIMARY PLASTICIZER

Has sufficient affinity to the polymer or resin so that it is considered compatible and therefore it may be used as the sole plasticizer.

PRINTED CIRCUIT

An electrical or electronic circuit produced mainly from copper clad laminates.

PRINTING OF PLASTICS

Methods of printing plastics materials, particularly thermoplastic film and sheet, have developed side by side with the growth of usage of the materials, and are today an important part of finishing techniques. Basically, the printing processes used are the same as in other industries, but the adaptation of machinery and development of special inks have been a constant necessity, particularly as new plastics materials have arrived, each with its own problems of surface decoration. Among the printing processes commonly used are gravure, flexographic, inlay (or valley) and silk screen.

PROGRAMMING

The extrusion of a parison which differs in thickness in the length direction in order to equalize wall thickness of the blown container. It can be done with a pneumatic or hydraulic device which actives the mandrel shaft and adjusts the mandrel position during parison extrusion (parison programmer, controller, or variator). It can also be done by varying extrusion speed on accumulator-type blow molding machines.

PROMOTER

A chemical, itself a feeble catalyst, that greatly increases the activitiy of a given catalyst.

PROTOTYPE MOLD

A simplified mold construction often made from a light metal casting alloy or from an epoxy resin in order to obtain information for the final mold and/or part design.

PULP

A form of cellulose obtained from wood or other vegetable matter by prolonged cooking with chemicals.

PULP MOLDING

Process by which a resin-impregnated pulp material is preformed by application of a vacuum and subsequently oven cured or molded.

PURGING

Cleaning one color or type of material from the cylinder of an injection molding machine or extruder by forcing it out with the new color or material to be used in subsequent production. Purging materials are also available.

PUSH UP

The bottom contour of a plastic container designed in such a manner as to allow an even bearing surface on the outside edge and to prevent the bottle from rocking.

PV VALUE

The maximum combination of pressure and velocity at which a material will operate continuously without lubrication.

PYROLYSIS

The chemical decomposition by the action of heat.

Q

QUENCH (Thermoplastics)

A process of shock cooling thermoplastic materials from the molten state.

QUENCH BATH

The cooling medium used to quench molten thermplastic materials to the solid state.

QUENCH-TANK EXTRUSION

The extruded film is cooled in a quench-water bath.

R

RADIO FREQUENCY (R.F.) PREHEATING

A method of preheating used for molding materials to facilitate the molding operation or reduce the molding cycle. The frequencies most commonly used are between 10 and 100 Mc./sec.

RADIO FREQUENCY WELDING

A method of welding thermoplastics using a radio frequency field to apply the necessary heat. Also known as high frequency welding.

RAM

See *Force Plug.*

RAM TRAVEL

The distance the injection ram moves in filling the mold, in either injection or transfer molding.

RAYON

The generic term for fibers, staple, and continuous filament yarns composed of regenerated cellulose, q.v., but also frequently used to describe fibers obtained from cellulose acetate or cellulose triacetate. Rayon fibers are similar in chemical structure to natural cellulose fibers (e.g., cotton) except that the synthetic fiber contains shorter polymer units. Most rayon is made by the viscose process.

REAM

Usually 500 sheets, 24" x 36" of industrial paper. Sometimes expressed as 3000 sq. ft.

RECESSED PANEL

A container design in which the flat area for labeling is indented or recessed. Also, see *Label Panel.*

RECIPROCATING SCREW

An extruder system in which the screw when rotating is pushed backwards by the molten polymer which collects in front of the screw. When sufficient material has been collected, the screw moves forward and forces the material through the head and die at a high speed.

RECTANGULAR

A particular shape of a container which has right angles with adjacent sides of unequal dimensions.

RECYCLED PLASTIC

A plastic prepared from used articles which have been cleaned and reground.

REFORMULATED PLASTIC

Recycled plastic that has been upgraded to alter or improve performance capability or to change characteristics through use of plasticizers, fillers, stabilizers, pigments, etc.

REGENERATED CELLULOSE (CELLOPHANE)

A transparent cellulose plastics material made by mixing cellulose xanthate with a dilute sodium hydroxide solution to form a viscose. Regeneration is carried out by extruding the viscose, in sheet form, into an acid bath to create regenerated cellulose. The material is very widely used as a packaging and overwrapping material of exceptional clarity. The film also has good electrical properties and is resistant to oils and greases. Included among recent application is the use of the material as a release agent in reinforced plastics moldings.

REGRIND

See *Reworked Plastic.*

REINFORCED MOLDING COMPOUND

Compound supplied by raw material produced in the form of ready-to-use materials; as distinguished from premix q.v.

REINFORCED PLASTICS

A plastic with high strength fillers imbedded in the composition, resulting in some mechanical properties superior to those of the base resin. (See also *Filler.*)

REINFORCEMENT

A strong inert material bound into a plastic to improve its strength, stiffness, and impact resistance. Reinforcements are usually long fibers of glass, sisal, cotton, etc. — in woven or nonwoven form. To be effective, the reinforcing material must form a strong adhesive bond with the resin.

RELATIVE HUMIDITY

Ratio of the quantity of water vapor present in the air to the quantity which would saturate it at any given temperature.

RELATIVE VISCOSITY

The relative viscosity of a polymer in solution is the ratio of the absolute viscosities of the solution (of stated concentration) and of the pure solvent at the same temperature.

$$n_r = \frac{n}{n_o}$$

where n = relative viscosity, n = absolute viscosity of polymer solution, n = absolute viscosity of pure solvent.

RELEASE AGENT

See *Parting Agent.*

RELIEF ANGLE

The angle of the cutaway portion of the pinch-off blade measured from a line parallel to the pinch-off land.

REPROCESSED PLASTIC

A thermoplastic prepared from scrap industrial plastic by other than the original processor.

RESILIENCY

Ability to quickly regain an original shape after being strained or distorted.

RESIN

Any of a class of solid or semi-solid organic products of natural or synthetic origin, generally of high molecular weight with no definite melting point. Most resins are polymers q.v.

RESINOID

Any of the class of thermosetting synthetic resins, either in their initial temporarily fusible state or in their final infusible state. Compare with *Thermoset.*

RESIN POCKET

An apparent accumulation of excess resin in a small, localized section visible on cut edges of molded surfaces.

RESISTIVITY

The ability of a material to resist passage of electrical current either through its bulk or on a surface. The unit of volume resistivity is the ohm-cm., of surface resistivity, the ohm.

RESTRICTED GATE

A very small orifice between runner and cavity in an injection or transfer mold. When the piece is ejected, this gate breaks cleanly, simplifying separation of runner from piece.

RESTRICTOR RING

A ring-shaped part protruding from the torpedo surface which provides increase of pressure in the mold to improve, e.g., welding of two streams.

RETAINER PLATE

The plate on which demountable pieces, such as mold cavities, ejector pins, guide pins, and bushings are mounted during molding; usually drilled for steam or water.

RETARDER

See *Inhibitor*.

REVERSE-ROLL COATING

The coating is premetered between rolls and then wiped off on the web. The amount of coating is controlled by the metering gap and also by the speed of rotation of the coating roll.

REWORKED PLASTIC

A thermoplastic from a processor's own production that has been reground or pelletized after having been previously processed by molding, extrusion, etc.

RHEOLOGY

Study of the deformation and flow of matter in terms of stress, strain and time.

RIB

A reinforcing member of a fabricated or molded part.

RIGID PLASTICS

For purposes of general classification, a plastic that has a modulus of elasticity either in flexure or in tension greater than 100,000 psi at 23°C and 50% relative humidity when tested in accordance with ASTM Methods D 747 or D 790 Test for stiffness of plastics.

RIGID PVC

Polyvinyl chloride or a polyvinyl chloride/acetate copolymer characterized by a relatively high degree of hardness; it may be formulated with or without a small percentage of plasticizer.

RIGID RESIN

One having a modulus high enough to be of practical importance, e.g., 10,000 psi or greater.

ROCKER

A plastic container with a bulged or deformed bottom, causing rocking of the container in the upright position.

ROCKWELL HARDNESS

A common method of testing a plastics material for resistance to indentation in which a diamond or steel ball, under pressure, is used to pierce the test specimen. The load used is expressed in kilograms and a 10-kilogram weight is first applied and the degree of penetration noted. The so-called major load (60 to 150 kilograms) is next applied and a second reading obtained. The hardness is then calculated as the difference between the two loads and expressed with nine different prefix letters to denote the type of penetrator used and the weight applied as the major load.

ROLLER COATING

Used for applying paints to raised designs or letters.

ROLL MILL

Two rolls placed in close relationship to one another used to admix a plastic material with other substances. The rolls turn at different speeds to produce a shearing action to the materials being compounded.

ROSIN
The hard resin, amber to black in color, left after the distillation of turpentine.

ROTATING SPREADER
A type of injection torpedo which consists of a finned torpedo which is rotated by a shaft extending through a tubular cross-section injection ram behind it.

ROTATIONAL CASTING
(or molding)
A method used to make hollow articles from plastisols and latices. Plastisol is charged into hollow mold capable of being rotated in one or two planes. The hot mold fuses the plastisol into a gel after the rotation has caused it to cover all surfaces. The mold is then chilled and the product stripped out.

ROUND SQUARE
Particular shape of a container which has sides of equal width with well-rounded corners and shoulders.

ROVING
A form of fibrous glass in which spun strands are woven into a tubular rope. The number of strands is variable but 60 is usual. Chopped roving is commonly used in preforming.

RUBBER
An elastomer capable of rapid elastic recovery after being stretched to at least twice its length at temperatures from 0 to 150°F, at any humidity. Specifically, Hevea or natural rubber, the standard of comparison for elastomers.

RUNNER
In an injection or transfer mold, the channel, usually circular, that connects the sprue with the gate to the cavity.

S

SAG
The extension locally (often near the die face) of the parison during extrusion by gravitational forces. This causes necking-down of the parison. Also refers to the flow of a molten sheet in a thermoforming operation.

SAMPLE
A small part or portion of a material or product intended to be representative of the whole.

SANDBLASTED
See *Grit Blasted.*

SANDWICH CONSTRUCTIONS
Panels composed of a lightweight core material — honeycomb, foamed plastic, etc., q.v. — to which two relatively thin, dense, high strength faces or skins are adhered.

SANDWICH HEATING
A method of heating a thermoplastic sheet prior to forming which consists of heating both sides of the sheet simultaneously.

SARAN PLASTIC
See *Vinylidene Chloride Plastics.*

SATURATED COMPOUNDS
Organic compounds which do not contain double or triple bonds and thus cannot add on elements or compounds.

SCAR
A characteristic mark on plastic containers which is confined mostly to the bottom. It is caused by the pinch-off operation and is often referred to as the length of the pinch-off.

SCRAP
Any product of a molding operation that is not part of the primary product. In compression molding, this includes flash, culls,

runners, and is not reusable as a molding compound. Injection molding and extrusion scrap (runners, rejected parts, sprues, etc.) can usually be reground and remolded.

SCREW PLASTICATING INJECTION MOLDING

A technique in which the plastic is converted from pellets to a viscose melt by means of an extruder screw which is an integral part of the molding machine. Machines are either single stage (in which plastication and injection are done by the same cylinder) or double stage in which the material is plasticated in one cylinder and then fed to a second for injection into a mold.

SDR

Standard Dimension Ratio for Plastic Pipe. The ratio of O.D. or I.D. to wall thickness.

SEALING PLANE

The plane on the inside of a bottle cap along the sealing surface.

SEALING SURFACE

The surface of the finish of the container on which the closure forms the seal.

SECONDARY PLASTICIZER (or Extender Plasticizer)

Has insufficient affinity for the resin to be compatible as the sole plasticizer and must be blended with a primary plasticizer. The secondary acts as a diluent with respect to the primary and the primary-secondary blend has less affinity for the resin than does the primary alone.

SEGREGATION

A close succession of parallel, relatively narrow and sharply defined, wavy lines of color on the surface of a plastic which differ in shade from surrounding areas, and create the impression that the components have separated.

SEIZING

Act of holding or grasping suddenly or forcibly.

SELF-EXTINGUISHING

A somewhat loosely-used term describing the ability of a material to cease burning once the source of flame has been removed.

SEMI-AUTOMATIC MOLDING MACHINE

A molding machine in which only part of the operation is controlled by the direct action of a human. The automatic part of the operation is controlled by the machine according to a predetermined program.

SEMIPOSITIVE MOLD

A mold which allows a small amount of excess material to escape when it is closed. See *Flash Mold, Positive Mold.*

SEMIRIGID PLASTIC

For purposes of general classification, a plastic that has a modulus of elasticity either in flexure or in tension of between 10,000 and 100,000 psi at $23^{\circ}C$ and 50% relative humidity when tested in accordance with ASTM Method D747 or D790 Test for Stiffness of Plastics.

SET

To convert a liquid resin or adhesive into a solid state by curing q.v., or by evaporation of solvent or suspending medium or by gelling.

SETTING TEMPERATURE

The temperature to which a liquid resin, an adhesive, or products or assemblies involving either is subjected to set the resin or adhesive.

SETTING TIME

The period of time during which a molded or extruded product, an assembly, etc., is subjected to heat and/or pressure to set the resin or adhesive.

SHARK SKIN

A surface irregularity of a container in the form of finely-spaced sharp ridges caused by a relaxation effect of the melt at the die exit.

SHEAR RATE

The overall velocity over the cross section of a channel with which molten polymer layers are gliding along each other or along the wall in laminar flow.

$$\text{shear rate} = \frac{\text{velocity}}{\text{clearance}} = \frac{\text{cm/sec.}}{\text{cm}} = \text{sec.} - 1.$$

SHEAR STRENGTH

(a) The ability of a material to withstand shear stress. (b) The stress at which a material fails in shear.

SHEAR STRESS

The stress developing in a polymer melt when the layers in a cross section are gliding along each other or along the wall of the channel (in laminar flow).

$$\text{shear stress} = \frac{\text{force}}{\text{area sheared}} = \text{psi}$$

SHEET (Thermoplastic)

A flat section of a thermoplastic resin with the length considerably greater than the width and 10 mils or greater in thickness.

SHEETER LINES

Parallel scratches or projecting ridges distributed over a considerable area of a plastic sheet.

SHEET TRAIN

The entire assembly necessary to produce sheet which includes extruder, die, polish rolls, conveyor, draw rolls, cutter and stacker.

SHELF LIFE

See *Storage Life.*

SHORE HARDNESS

A method of determining the hardness of a plastic material using a scelroscope. This device consists of a small conical hammer fitted with a diamond point and acting in a glass tube. The hammer is made to strike the material under test and the degree of rebound is noted on a graduated scale. Generally, the harder the material the greater will be the rebound.

SHORT or SHORT SHOT

In injection molding, failure to fill the mold completely.

SHOT

The yield from one complete molding cycle, including scrap.

SHOT CAPACITY

The maximum weight of material which an accumulator can push out with one forward stroke of the ram.

SHRINKAGE

Contraction of a container upon cooling.

SHRINK FIXTURE

See *Cooling Fixture.*

SHRINK MARK

An imperfection, a depression in the surface of a molded material where it has retracted from the mold.

SHRINK WRAPPING

A technique of packaging in which the strains in a plastics film are released by raising the temperature of the film thus causing it to shrink over the package. These shrink characteristics are built into the film during its manufacture by stretching it under controlled temperatures to produce orientation, q.v., of the molecules. Upon cooling, the film retains its stretched condition, but reverts toward its original dimensions when it is heated. Shrink film gives good protection to the products packaged and has excellent clarity.

SIAMESE BLOW

A colloquial term applied to the technique of blowing two or more parts of a product in a single blow and then cutting them apart.

SIDE BARS

Loose pieces used to carry one or more molding pins, and operated from outside the mold.

SIDE DRAW PINS

Projections used to core a hole in a direction other than the line of closing of a mold, and which must be withdrawn before the part is ejected from the mold.

SILK SCREEN PRINTING
(Screen process decorating)

This printing method, in its basic form, involves laying a pattern of an insoluble material, in outline, on a finely woven fabric, so that when ink is drawn across it, it is able to pass through the screen only in the desired areas.

SINGLE CAVITY MOLD
(Injection)

An injection mold having only one cavity in the body of the mold, as opposed to a multiple cavity mold or family mold which have numerous cavities.

SILICONE

One of the family of polymeric materials in which the recurring chemical group contains silicon and oxygen atoms as links in the main chain. At present these compounds are derived from silica (sand) and methyl chloride. The various forms obtainable are characterized by their resistance to heat.

Silicones are used in the following applications: (a) Greases for lubrication. (b) Rubber-like sheeting for gaskets, etc.(c) Heat-stable fluids and compounds for waterproofing, insulating, etc. (d) Thermosetting insulating varnishes and resins for both coating and laminating.

SINK MARK

A shallow depression or dimple on the surface of an injection molded part due to collapsing of the surface following local internal shrinkage after the gate seals. May also be an incipient short shot.

SINKING A MOLD

See *Hobbing*.

SINTERING

The partial welding together of powder particles at temperature near the melting point.

SI UNITS

The International System of Units (Systems International) is a modernized version of the metric system established by international agreement. It provides a logical and interconnected framework for all measurements in science, industry and commerce. Officially abbreviated SI, the system is built upon a foundation of seven base units.

SIZING

(*n.*) The process of applying a material to a surface to fill pores and thus reduce the absorption of the subsequently applied adhesive or coating or to otherwise modify the surface. Also, the surface treatment applied to glass fibers used in reinforced plastics. The material used is sometimes called Size.

SKIN

A relatively dense layer at the surface of a cellular material.

SLIP ADDITIVE

A modifier that acts as an internal lubricant which exudes to the surface of the plastic during and immediately after processing. In other words, a non-visible coating blooms to the surface to provide the necessary lubricity to reduce coefficient of friction and thereby improve slip characteristics.

SLIP FORMING

Sheet forming technique in which some of the plastic sheet material is allowed to slip through the mechanically operated clamping rings during a stretch-forming operation.

SLIP-PLANE

Plane within transparent material visible in reflected light, due to poor welding and shrinkage on cooling.

SLOT EXTRUSION

A method of extruding film sheet in which the molten thermoplastic compound is forced through a straight slot.

SLURRY PREFORMING

Method of preparing reinforced plastics preforms by wet processing techniques similar to those used in the pulp molding q.v., industry.

SLUSH MOLDING

Method for casting thermoplastics, in which the resin in liquid form is poured into a hot mold where a viscous skin forms. The excess slush is drained off, the mold is cooled, and the molding stripped out.

SNAP-BACK FORMING

Sheet forming technique in which an extended heated plastic sheet is allowed to contract over a male form shaped to the desired contours.

SOLID LUBRICANT

See *Lubricant, Solid.*

SOLUTION

Homogeneous mixture of two or more componets, such as a gas dissolved in a gas or liquid, or a solid in a liquid.

SOLVENT

Any substance, usually a liquid which dissolves other substances.

SOLVENT MOLDING

Process for forming thermoplastic articles by dipping a male mold in a solution or dispersion of the resin and drawing off the solvent to leave a layer of plastic film adhering to the mold.

SPACER CABLE

A system of 2-15kv primary power distribution in which three partially insulated or covered phase wires and a high-strength messenger-ground wire are mounted in plastic or ceramic insulating spacers.

SPANISHING

A method of depositing ink in the valleys of embossed plastics film.

SPECIFIC GRAVITY

The density (mass per unit volume) of any material divided by that of water at a standard temperature, usually $4^{\circ}C$. Since water's density is nearly 1.00 g./cc., density in g./cc. and specific gravity are numerically nearly equal.

SPECIFIC HEAT

The amount of heat required to raise a specified mass by one unit of a specified temperature.

SPECIFIC VISCOSITY

The specific viscosity of a polymer is the relative viscosity of a solution of known concentration of the polymer minus one. It is usually determined for a low concentration of the polymer (0.5 g. per 100 ml. of solution or less).

$$n\text{SP} = \frac{n - n^o}{n^o} = n^r - 1,$$

where $n\text{SP}$ = specific viscosity, n^r = relative viscosity.

SPECIMEN
A piece or portion of a sample used to make a test.

SPECULAR TRANSMITTANCE
The transmittance value obtained when the measured transmitted flux includes only that transmitted in essentially the same direction as the incident flux.

SPIDER
(1) In a molding press, that part of an ejector mechanism which operates the ejector pins. (2) In extrusion, a term used to denote the membranes supporting a mandrel within the head/die assembly.

SPIDER LINES
Vertical marks on the parison (container) caused by improper welding of several melt flow fronts formed by the legs with which the torpedo is fixed in the extruder head.

SPINNERET
A type of extrusion die, i.e., a metal plate with many tiny holes, through which a plastic melt is forced to make fine fibers and filaments. Filaments may be hardened by cooling in air, water, etc., or by chemical action.

SPINNING
Process of making fibers by forcing plastic melt through spinneret.

SPIN WELDING
A process of fusing two objects together by forcing them together while one of the pair is spinning, until frictional heat melts the interface. Spinning is then stopped and pressure held until they are frozen together.

SPIRAL FLOW TEST
A method for determining the flow properties of a thermoplastic resin in which the resin flows along the path of a spiral cavity. The length of the material which flows into the cavity and its weight gives a relative indication of the flow properties of the resin.

SPIRAL MOLD COOLING
A method of cooling injection molds or similar molds wherein the cooling medium flows through a spiral cavity in the body of the mold. In injection molds, the cooling medium is introduced at the center of the spiral, near the sprue section, as more heat is localized in this section.

SPLIT-RING MOLD
A mold in which a split cavity block is assembled in a chase to permit the forming of undercuts in a molded piece. These parts are ejected from the mold and then separated from the piece.

SPRAY COATING
Usually accomplished on continuous webs by a set of reciprocating spray nozzles traveling laterally across the web as it moves.

SPRAYED METAL MOLDS
Mold made by spraying molten metal onto a master until a shell of predetermined thickness is achieved. Shell is then removed and backed up with plaster, cement, casting resin, or other suitable material. Used primarily as a mold in sheet-forming processes.

SPRAY-UP
Covers a number of techniques in which a spray gun is used as the

processing tool. In reinforced plastics, for example, fibrous glass and resin can be simultaneously deposited in a mold. In essence, roving is fed through a chopper and ejected into a resin stream which is directed at the mold by either of two spray systems. In foamed plastics, very fast-reacting urethane foams or epoxy foams are fed in liquid streams to the gun and sprayed on the surface. On contact, the liquid starts to foam.

SPREADER
A streamlined metal block placed in the path of flow of the plastics material in the heating cylinder of extruders and injection molding machines to spread it into thin layers, thus forcing it into intimate contact with the heating areas.

SPRUE
Feed opening provided in the injection or transfer mold; also the slug formed at this hole. Spur is a shop term for the sprue slug.

SPRUE BUSHING
A hardened steel insert in an injection mold which contains the tapered sprue hole and has a suitable seat for the nozzle of the injection cylinder. Sometimes called an Adapter.

SPRUE GATE
A passageway through which molten resin flows from the nozzle to the mold cavity.

SPRUE LOCK
In injection molding, a portion of the plastic composition which is held in the cold slug well by an undercut; used to pull the sprue out of the bushing as the mold is opened. The sprue lock itself is pushed out of the mold by an ejector pin. When the undercut

occurs on the cavity block retainer plate, this pin is called the Sprue Ejector Pin.

STABILIZER
An ingredient used in the formulation of some plastics, especially elastomers, to assist in maintaining the physical and chemical properties of the compounded materials at their initial values throughout the processing and service life of the material.

STAPLE
Refers to textile fibers of a short length, usually ½ to 3", for natural fibers and sometimes larger for synthetics.

STATIONARY PLATEN
The large front plate of an injection molding machine to which the front plate of the mold is secured during operation. This platen does not move during normal operations.

STEAM MOLDING
(expandable polystyrene)
Used to mold parts from pre-expanded beads of polystyrene using steam as a source of heat to expand the blowing agent in the material. The steam in most cases is contacted intimately with the beads directly or may be used indirectly to heat mold surfaces which are in contact with the beads.

STEAM PLATE
Mounting plate for molds, cored for circulation of steam.

STEREOSPECIFIC PLASTICS
Implies a specific or definite order of arrangement of molecules in space. This ordered regularity of the molecules in contrast to the branched or random arrangement found in other plastics permits close packing of the molecules and leads to high crystallinity (i.e., as in polypropylene).

STITCHING

The progressive welding of thermoplastic materials by successive applications of two small mechanically operated electrodes, connected to the output terminals of a radio frequency generator, using a mechanism similar to that of a normal sewing machine.

STOCK TEMPERATURE

See *Melt Temperature.*

STORAGE LIFE

The period of time during which a liquid resin or packaged adhesive can be stored under specified temperature conditions and remain suitable for use. Storage life is sometimes called *Shelf Life.*

STRAIGHT-SIDED ROUND

A round bottle with straight side walls from shoulder to base.

STRESS CRACK

An external or internal crack in a plastic caused by tensile stresses less than its short-time mechanical strength.

STRETCH FORMING

A plastic sheet forming technique in which the heated thermoplastic sheet is stretched over a mold and subsequently cooled.

STRIATION

Rippling of thick parisons, caused by a local orientation effect in the melt by the spider legs.

STRIPPER-PLATE

A plate that strips a molded piece from core pins or force plugs. The stripper-plate is set into operation by the opening of the mold.

SUBMARINE GATE

A type of edge gate where the opening from the runner into the mold is located below the parting line or mold surface as opposed to conventional edge gating where the opening is machined into the surface of the mold. With submarine gates, the item is broken from the runner system on ejection from the mold.

SURFACE RESISTIVITY

The electrical resistance between opposite edges of a unit square of insulating material. It is commonly expressed in ohms. (Also covered in ASTM D257-54T.)

SURFACE TREATING

Any method of treating a polyolefin so as to alter the surface and render it receptive to inks, paints, lacquers, and adhesives such as chemical, flame, and electronic treating.

SURFACTANT

A compound that affects interfacial tensions between two liquids. It usually reduces surface tension.

SURGING

Unstable pressure build-up in an extruder leading to variable throughput and waviness of the parison.

SUSPENSION

A mixture of fine particles of any solid with a liquid or gas. The particles are called the disperse phase, the suspending medium is called the continuous phase.

SWEATING

Exudation of small drops of liquid, usually a plasticizer or softener, on the surface of a plastic part.

SYNDIOTACTIC

A chain of molecules in which the methyl groups alternate regularly on opposite sides of the chain.

SYNERESIS

The contraction of a gel accompanied by the separation of a liquid.

SYNERGISM

A term used to describe the use of two or more stabilizers in an organic material where the combination of such stabilizers improves the stability to a greater extent than could be expected from the additive effect of each stabilizer.

SYNTACTIC FOAM
A material consisting of hollow sphere fillers in a resin matrix.

T

TAB GATED
A small removable tab of approximately the same thickness as the mold item, usually located perpendicular to the item. The tab is used as a site for edge gate location, usually on items with large flat areas.

TACK
Stickiness of an adhesive, measurable as the force required to separate an adherend from it by viscous or plastic flow of the adhesive.

TACK RANGE
The period of time in which an adhesive will remain in the tacky-dry condition after application to an adherend, under specified conditions of temperature and humidity.

TAPERED CYLINDER
Refers to a particular shape of a container in which the circular cross section at the top is smaller in diameter than that at the bottom, or vice versa.

TAPPING
Cutting threads in the walls of a circular hole.

T-DIE
A term used to denote a center-fed, slot extrusion die for film which in combination with the die adapter resembles an inverted T.

TELOMER
A polymer composed of molecules having terminal groups incapable of reacting with additional monomers, under the conditions of the synthesis, to form larger polymer molecules of the same chemical type.

TENACITY (gpd)
The term generally used in yarn manufacture and textile engineering to denote the strength of a yarn or of a filament for its given size. Numerically it is the grams of breaking force per denier unit of yarn or filament size; grams per denier, gpd. The yarn is usually pulled at the rate of 12 in./min. Tenacity equals breaking strength (gms) divided by denier. (Tenacity, gpd) (Spec. Gravity) 12,800 = (Tensile Strength, psi).

TENSILE BAR (Specimen)
A compression or injection molded specimen of specified dimensions which is used to determine the tensile properties of a material.

TENSILE STRENGTH
The pulling stress, in psi, required to break a given specimen. Area used in computing strength is usually the original, rather than the necked-down area.

THERIMAGE
A trademark for a decorating process for plastic which transfers the image of a label or decoration to the object under the influence of heat and light pressure.

THERMAL CONDUCTIVITY
Ability of a material to conduct heat; physical constant for quantity of heat that passes through unit cube of a substance in unit of time when difference in temperature of two faces is 1°.

THERMAL DEGRADATION

Deterioration by heat.

THERMAL EXPANSION
(Coefficient of)

The fractional change in length (sometimes volume, specified) of a material for a unit change in temperature. Values for plastics range from 0.01 to 0.2 mils/in., $^{\circ}$C.

THERMAL STRESS CRACKING (TSC)

Crazing and cracking of some thermoplastic resins which results from over-exposure to elevated temperatures.

THERMOFORMING

Any process of forming thermoplastic sheet which consists of heating the sheet and pulling it down onto a mold surface.

THERMOFORMS

The product which results from a thermoforming operation.

THERMOPLASTIC

(a.) Capable of being repeatedly softened by heat and hardened by cooling (n.) — A material that will repeatedly soften when heated and harden when cooled. Typical of the thermoplastics family are the styrene polymers and copolymers, acrylics, cellulosics, polyethylenes, vinyls, nylons, and the various fluorocarbon materials.

THERMOSET

A material that will undergo or has undergone a chemical reaction by the action of heat, catalysts, ultra-violet light, etc., leading to a relatively infusible state. Typical of the plastics in the thermosetting family are the aminos (melamine and urea), most polyesters, alkyds, epoxies, and phenolics.

THIXOTROPIC

Said of materials that are gel-like at rest but fluid when agitated. Liquids containing suspended solids are apt to be thixotropic. Thixotropy is desirable in paints.

THREAD CONTOUR

The shape or type of thread design as observed in a cross section along the major axis, i.e., flatheaded, square, round, etc.

THREAD PLUG

A part of a mold that shapes an internal thread and must be unscrewed from the finished piece.

TIE BARS

Bars which provide structural rigidity to the clamping mechanism often used to guide platen movement.

TOGGLE ACTION

A mechanism which exerts pressure developed by the application of force on a knee joint. It is used as a method of closing presses and also serves to apply pressure at the same time.

TOLERANCE

A specified allowance for deviations in weighing, measuring, etc., or for deviations from the standard dimensions or weight.

TOP BLOW

A specific type of blow molding machine which forms hollow articles by injecting the blowing into the parison at the top of the mold.

TORPEDO (or Spreader)

A streamlined metal block placed in the path of flow of the plastics materials in the heating cylinder of extruders and injection molding machines to spread it into thin layers, thus forcing it into intimate contact with heating areas.

TORSION

Stress caused by twisting a material.

TRACKING

A phenomenon wherein a high voltage source current creates a leakage or fault path across the surface of an insulating material by slowly but steadily forming a carbonized path.

TRANSFER MOLDING

A method of forming articles by fusing a plastic material in a chamber and then forcing essentially the whole mass into a hot mold where it solidifies.

TRANSLUCENT

Descriptive of a material or substance capable of transmitting some light, but not clear enough to be seen through.

TRANSPARENT

Descriptive of a material or substance capable of a high degree of light transmission e.g., glass. Some polypropylene films and acrylic moldings are outstanding in this respect.

TREE WIRE

A special type of power line wire designed for installation in wooded areas.

TUMBLING

Finishing operation for small plastic articles by which gates, flash, and fins are removed and/or surfaces are polished by rotating them in a barrel together with wooden pegs, sawdust, and polishing compounds.

TURNING TABLE

A rotating table or wheel carrying various molds in a multi-mold single parison blow molding operation.

U

ULTIMATE STRENGTH

Term used to describe the maximum unit stress a material will withstand when subjected to an applied load in a compression, tension, or shear test.

ULTRASONIC SEALING

A film sealing method in which sealing is accomplished through the application of vibratory mechanical pressure at ultrasonic frequencies (20 to 40 kc.). Electrical energy is converted to ultrasonic vibrations through the use of either a magnetostrictive or piezoelectric transducer. The vibratory pressures at the film interface in the sealing area develop localized heat losses which melt the plastic surfaces effecting the seal.

ULTRAVIOLET

Zone of invisible radiations beyond the violet end of the spectrum of visible radiations. Since UV wavelengths are shorter than the visible, their photons have more energy, enough to initiate some chemical reactions and to degrade most plastics.

UNDERCUT

(a.) Having a protuberance or indentation that impedes withdrawal from a two-piece, rigid mold. Flexible materials can be ejected intact even with slight undercuts. (n.) Any such protuberance or indentation; depends also on design of mold.

UNICELLULAR

With foamed plastics, each cell an isolated unit. Equals "closed-cell."

UNIT MOLD

A simple mold which comprises only a single cavity without further mold devices to be used for the production of sample containers having a shape which is difficult to blow.

UNSATURATED COMPOUNDS

Any compound having more than one bond between two adjacent atoms, usually carbon atom, and capable of adding other atoms at that point to reduce it to a single bond.

UREA PLASTICS

Plastics based on resins made by the condensation of urea and aldehydes.

URETHANE

See description under *Isocyanate Resins.*

UV STABILIZER (Ultraviolet)

Any chemical compound which, when admixed with a thermoplastic resin, selectively absorbs UV rays.

V

VACUUM FORMING

Method of sheet forming in which the plastic sheet is clamped in a stationary frame, heated, and drawn down by a vacuum into a mold. In a loose sense, it is sometimes used to refer to all sheet forming techniques, including *Drape Forming* q.v., involving the use of vacuum and stationary molds.

VACUUM METALIZING

Process in which surfaces are thinly coated with metal by exposing them to the vapor of metal that has been evaporated under vacuum (one millionth of normal atmospheric pressure).

VALLEY PRINTING

Ink is applied to the high points of an embossing roll and subsequently deposited in what becomes the valleys of the embossed plastic material.

VEHICLE

The liquid medium in which pigments, etc, are dispersed in coatings such as paint, q.v., and which enable the coating to be applied.

VENT

In a mold, a shallow channel or minute hole cut in the cavity to allow air to escape as the material enters.

VENTURI DISPERSION PLUG

A plate having an orifice with a conical relief drilled therein which is fitted in the nozzle of an injection molding machine to aid in the dispersion of colorants in a resin.

VERTICAL FLASH RING

The clearance between the force plug and the vertical wall of the cavity in a positive or semi-positive mold; also the ring of excess material which escapes from the cavity into this clearance space.

VINYL ACETATE PLASTICS

Plastics based on polymers of vinyl acetate or copolymers of vinyl acetate with other monomers, the vinyl acetate being in greatest amount by mass.

VINYL CHLORIDE PLASTICS

Plastics based on polymers of vinyl chloride or copolymers of vinyl chloride with other monomers, the vinyl chloride being in greatest amount by mass.

VINYLIDENE CHLORIDE PLASTICS

Plastics based on polymer resins made by the polymerization of vinylidene chloride or copolymerization of vinylidene chloride with other unsaturated compounds, the vinylidene chloride being in the greatest amount by weight.

VIRGIN MATERIAL

A plastic material in the form of pellets, granules, powder, flock, or liquid that has not been subjected to use or processing other than that required for its initial manufacture.

VISCOSITY

Internal friction or resistance to flow of a liquid. The constant ratio of shearing stress to rate of shear. In liquids for which this ratio is a function of stress, the term "apparent viscosity" is defined as this ratio.

VISCOSITY, INHERENT

The logarithmic viscosity number determined by dividing the natural logarithm of the relative viscosity (sometimes called viscosity ratio) by the concentration in grams per 100 mls. of solution.

VISCOSITY, RELATIVE
(or Viscosity Ratio)

Determined by dividing the average efflux time of the solution by the average efflux time of the pure solvent.

VOIDS

(1) In a solid plastic, an unfilled space of such size that it scatters radiant energy such as light.
(2) A cavity unintentionally formed in a cellular material and substantially larger than the characteristics individual cells.

VOLATILES

That portion of a substance that is readily vaporized.

VOLUME RESISTIVITY (Specific Insulation Resistance)

The electrical resistance between opposite faces of a 1-cm. cube of insulating material. It is measured under prescribed conditions using a direct current potential after a specified time of electrification. It is commonly expressed in ohm-centimeters. The recommended test is ASTM D257-54T.

VULCANIZATION

The chemical reaction which induces extensive changes in the physical properties of a rubber and which is brought about by reacting the rubber with sulphur and/or other suitable agents. The changes in physical properties include decreased plastic flow, reduced surface tackiness, increased elasticity, much greater tensile strength, and considerably less solubility. More recently, certain thermoplastics, e.g., polyethylene, have been formulated to the vulcanizable. Cross-linking is encouraged, thereby giving resistance to deformation of flow above the melting point.

W

WAIST

The central portion of a container which has a smaller cross section than the adjacent areas.

WARPAGE

Dimensional distortion in a plastic object after molding.

WEATHEROMETER

An instrument which is utilized to subject articles to accelerated weathering conditions, e.g., rich UV source and water spray.

WEB

A thin sheet in process in a machine. The molten web is that which issues from the die. The substrate web is the substrate being coated.

WELD LINES

A mark on a container caused by incomplete fusion of two streams of molten polymer. See *Spider Lines*.

WELD MARK (also Flow Line)

A mark on a molded plastics piece made by the meeting of two flow fronts during the molding operation.

WELDING

Joining thermoplastic pieces by one of several heat-softening processes. In hot-gas welding, the material is heated by a jet of hot air or inert gas directed from a welding "torch" onto the area of contact of the surfaces which are being welded. Welding operations to which this method is applied normally require the use of a filler rod. In *Spin-Welding* q.v., the heat is generated by friction. Welding also includes heat sealing and the terms are synonymous in some foreign countries including Britain.

WET STRENGTH

The strength of paper when saturated with water, especially used in discussions of processes whereby the strength of paper is increased by the addition, in manufacture, of plastics resins. Also, the strength of an adhesive joint determined immediately after removal from a liquid in which it has been immersed under specified conditions of time, temperature, and pressure.

WINDOW

A defect in a thermoplastics film, sheet or molding, caused by the incomplete "plasticization" of a piece of material during processing. It appears as a globule in an otherwise blended mass. See also *Fish Eye*.

WIRE TRAIN

The entire assembly which is utilized to produce a resin-coated wire which normally consists of an extruder, a crosshead and a die, cooling means, and feed and take-up spools for the wire.

WOOD MODEL

A model of a container made from wood to assist in the design of a container.

WORKING LIFE

The period of time during which a liquid resin or adhesive, after mixing with catalyst, solvent, or other compounding ingredients, remains usable.

WRINKLE

An imperfection in reinforced plastics that has the appearance of a wave molded into one or more plies of fabric or other reinforcing material.

Y

YOUNG'S MODULUS OF ELASTICITY

The modulus of elasticity in tension. The ratio of stress in a material subjected to deformation.

YIELD VALUE (Yield Strength)

The lowest stress at which a material undergoes plastic deformation. Below this stress, the material is elastic; above it, viscous.

Z

ZYZZOGETON

Not a plastics term but the last word in the dictionary.

Index